VEHICULAR TECHNOLOGIES - DEPLOYMENT AND APPLICATIONS

Edited by **Lorenzo Galati Giordano**
and **Luca Reggiani**

Vehicular Technologies - Deployment and Applications

http://dx.doi.org/10.5772/46112
Edited by Lorenzo Galati Giordano and Luca Reggiani

Contributors

Hui Zhang, Javier Prieto, Santiago Mazuelas, Alfonso Bahillo, Patricia Fernández, Rubén M. Lorenzo, Evaristo J. Abril, Anna Maria Vegni, Roberto Cusani, Mauro Biagi, Marcela Mejia, Ramiro Chaparro, Rausley Adriano Amaral De Souza, André Antônio Dos Anjos, Dayan Adionel Guimarães, Antonio F. Gómez-Skarmeta, José Santa Lozano, Pedro J. Fernandez Ruiz, Fernando Bernal, Lorenzo Galati Giordano, Luca Reggiani, Laura Dossi, Roberto Lambiase, Xiaodong Xu, Matthias Pätzold, Batool Talha

Published by InTech

Janeza Trdine 9, 51000 Rijeka, Croatia

Copyright © 2013 InTech

Notice

Publishing Process Manager Danijela Duric

Technical Editor InTech DTP team

Cover InTech Design team

Additional hard copies can be obtained from orders@intechopen.com

Vehicular Technologies - Deployment and Applications, Edited by Lorenzo Galati Giordano
and Luca Reggiani
p. cm.
ISBN 978-953-51-0992-1

For EU product safety concerns:
IN TECH d.o.o., Prolaz Marije Krucifikse Kozulić 3, 51000 Rijeka, Croatia,
info@intechopen.com or visit our website at intechopen.com.

Meet the editors

Lorenzo Galati Giordano received his PhD in Wireless Communications from Politecnico di Milano in 2008. During his PhD he has been a visiting Marie-Curie student at Centre for Wireless Network Design, University of Bedforshire (UK), supported by the European project RanPlan-HEC EU-FP6. In 2009 he has worked as a temporary researcher for the Italian National Research Council (IEIIT-CNR). Since 2010, he has moved to industry working as a consultant for Nokia Siemens Network (Germany) and as a 4G R&D Engineer for Azcom Technology (Italy). Lorenzo's main areas of interest are: 4G Architectures, Cross Layer Scheduling, Self-Organizing Networks, Vehicular and Mobile Networks, Small Cells deployment, PHY receiver design.

Luca Reggiani received the PhD in Electronics and Communications in 2001 from the Politecnico di Milano (Italy). As a consultant he has collaborated with several industries in the field of wireless communications and magnetic recording. Since 2003, he has been a researcher at Politecnico di Milano and, since 2002, he has been holding courses on Signal and Transmission Theory and Wireless Communication Systems. He has joined Italian and European research programs, especially in the field of cellular radio broadband systems and personal short range UWB (Ultra Wide Band) systems. His main publications and patents are in the physical layer of communications and radio transmission.

Contents

Preface

The application of communications technology to vehicles and road infrastructures is one of the most promising areas for the industrial and research investments in the next two decades. The communication capability among vehicles (i.e. vehicle-to-vehicle or V2V) and between each vehicle and the infrastructure (i.e. vehicle-to-road infrastructure or V2I) will potentially have a huge impact on traffic applications and emergency situations, making the future smart roads a safer and more comfortable experience w.r.t. nowadays conditions. In particular, vehicular communications, w.r.t. other systems, will have a relevant, crucial component constituted by machine-to-machine (M2M), automatic communications in addition to the active traditional data traffic requested and generated by the connected users. From this perspective, for the wide spectrum of applications and services and for their integration in the urban environment, smart roads are considered a key infrastructure in the context of wider scenarios as the smart cities and the smart communities.

The efficient, fast and flexible distribution of information in these scenarios requires an ensemble of advanced technologies in different fields, which respond to numerous challenging, technological issues; vehicular communications (i) require an efficient usage of the spectrum since they have to coexist with other communication systems especially in dense urban areas, (ii) work in a high mobility environment with potentially high user density in heavy traffic conditions and, at the same time, (iii) should guarantee high priority and reliable transmission for the high priority and critical emergency transmissions. In addition, the cost of the network infrastructure for V2I should be limited because of the necessity of covering thousands of kilometers of existing roads and highways in a pervasive way. Today the future smart roads services seem relatively possible in a near future, because all the equipments required for deploying the dedicated infrastructure are already available for mass production and they have a high degree of maturity. Nevertheless, the deployment of such a dedicated infrastructure is not trivial and does require further studies in order to cope with budget constraints and high reliability and capability.

Hopefully, this book will serve as a handbook for students, researchers and practitioners in the area of vehicular communications explaining and presenting solutions for some of the most critical issues in this field and inspiring new research directions. The book is organized in Sections, which respond to different layers and aspects of the vehicular technology. Each Section contains 2 or 3 chapters, focused on a certain aspect of the problem; within every chapter the reader will be given an overview of the background information on the subject and a description of the authors' original proposed solution. The chapters in each Section should give the reader a general idea where the current research efforts in vehicular technology are heading.

Section 1 is dedicated to a general introduction of vehicular technologies.

In Chapter 1, the authors present a precious review of the main communication protocols used in vehicular ad hoc networks and of the main technologies and applications, used to support emerging inter and intra-vehicle communications. The chapter provides an insight of the future smart vehicle, which will be considerably different from today's vehicle, due to its increased ability to control critical situations and take active decisions.

Section 2 is dedicated to the investigation of several issues concerning infrastructures and cells deployments, which are crucial for an effective implementation of V2I communications and its integration with the V2V part.

In Chapter 2, the authors present an efficient method for base station planning (sites and cells dimensions), characterized by the exploitation of stochastic geometry. The chapter introduces a general framework for base station design and it analyzes the frequency planning, investigating its theoretical basis from the perspectives of multigraph theory, algebraic analysis principle, extension theory and Stackelberg theory. Based on this theoretical analysis, it is presented a soft fractional frequency reuse scheme, which could have the flexibility and performance necessary for advanced vehicular communications infrastructures.

In Chapter 3, the authors have studied the deployment of LTE small base stations along roads characterized with high traffic density in order to provide vehicle-to-infrastructure (V2I) communication services. The chapter presents numerical results for addressing the V2I deployment and the set-up of the geometric configuration of the cells, distinguishing different classes of traffic in the system, from high priority, associated to emergency situations, to medium and low priority services like weather forecast, traffic situation and other multimedia services. An interesting characteristic of this deployment is also the use of the LTE relay capabilities for internal backhauling between adjacent base stations, a low cost approach suitable for V2I networks.

In Chapter 4, the authors present a networking stack that follows the current ESO/ETSI trends towards an intelligent transport system communication architecture. A system that supports different communication technologies for the vehicle connectivity has been defined and implemented and a secure IPv6 network mobility solution is also presented and tested in a realistic, experimental environment. The provided results indicate that the data rate required by most of the traffic services can be guaranteed and also the non-highly critical security services turn out to be compatible with the latency results.

Section 3 is dedicated to the analysis of several issues related to the access procedures in V2V and V2I communications.

In Chapter 5, the authors propose the Maximum Utility Principle for multiple access point cooperative communication suitable for vehicle environments and based on Dijkstra's Shortest Path Algorithm. This technique is applied in the access control and in the handover strategy in order to improve the performance of mobile communication in high speed mobility scenario. It is shown that proper combinations of scheduling and access control strategies, as those proposed in the chapter, enhance performance in terms of system throughput and user fairness in different conditions of system load and access users.

In Chapter 6, the authors present the review of trust and reputation models for V2V networks from different technical, technological and performance perspectives. The achievement of trustworthiness among peers has some strong challenges due to vehicles' high speeds and channel saturation in urban streets or suburbs highways during specific time periods. Some novel approaches and innovative models are proposed for facing these challenges and tackling some of drawbacks and downsides of current solutions.

Finally, **Section 4** is dedicated to three advanced services that will become essential part of the future, most critical applications of V2V and V2I communications: localization, spectrum sensing, and relay-based cooperative networks.

In Chapter 7, the authors present an adaptive likelihood function for robust data fusion in localization systems and an effective method for range and position estimation. This method has some noticeable advantages over conventional techniques, which result in an important performance improvement, corroborated by experimental and simulation results under NLOS and multipath conditions. The chapter provides a valuable contribution to the selection of appropriate techniques for localization in V2I communication systems and the proposed technique can be extended also to cooperative localization in the context of V2V communications.

In Chapter 8, the authors discuss about the spectrum sensing for the detection of underutilized frequency bands in the context of cognitive radio (CR) applications for vehicular networks. After a review of the main concepts related to cooperative spectrum sensing, with focus on those detection techniques based on the eigenvalues of the received signal covariance matrix, a realistic sensing environment model is described and developed in a MATLAB-based simulation platform. This contribution allows a meaningful assessment of the performance of the spectrum sensing techniques that can be applied in a cognitive radio scenario.

In Chapter 9, the authors analyze the statistical properties of mobile-to-mobile (M2M) fading channels with equal gain combining (EGC) under line-of-sight propagation conditions and relaying. This is a propagation scenario that is particularly interesting for V2V and V2I communications since the transmission can occur through the use of ad-hoc relays between the mobile transmitter and the final destination. In this work, simple and closed-form analytical approximations are derived for the probability density function, the cumulative distribution function, the level-crossing rate and the average duration of fades of a dual-hop amplify-and-forward relay network.

Lorenzo Galati Giordano,
Azcom Technology s.r.l.,
Rozzano, Milano, Italy

Luca Reggiani,
Dipartimento di Elettronica e Informazione,
Politecnico di Milano, Milano, Italy

Section 1

Introduction to Vehicular Technologies

Chapter 1

Smart Vehicles, Technologies and Main Applications in Vehicular Ad hoc Networks

Anna Maria Vegni, Mauro Biagi and Roberto Cusani

Additional information is available at the end of the chapter

http://dx.doi.org/10.5772/55492

1. Introduction

Vehicular Ad hoc NETworks (VANETs) belong to a subcategory of traditional Mobile Ad hoc NETworks (MANETs). The main feature of VANETs is that mobile nodes are vehicles endowed with sophisticated "on-board" equipments, traveling on constrained paths (*i.e.*, roads and lanes), and communicating each other for message exchange via Vehicle-to-Vehicle (V2V) communication protocols, as well as between vehicles and fixed road-side Access Points (*i.e.*, wireless and cellular network infrastructure), in case of Vehicle-to-Infrastructure (V2I) communications [1].

Future networked vehicles represent the future convergence of *computers, communications infrastructure,* and *automobiles* [2]. Vehicular communication is considered as an enabler for driverless cars of the future. Presently, there is a strong need to enable vehicular communication for applications such as safety messaging, traffic and congestion monitoring and general purpose Internet access.

VANET is a term used to describe the spontaneous ad hoc network formed over vehicles moving on the roadway. Vehicular networks are fast emerging for developing and deploying new and traditional applications. More in detail, VANETs are characterized by high mobility, rapidly changing topology, and ephemeral, one-time interactions. Basically, both VANETs and MANETs are characterized by the movement and self-organization of the nodes (*i.e.*, vehicles in the case of VANETs). However, due to driver behavior, and high speeds, VANETs characteristics are fundamentally different from typical MANETs. VANETs are characterized by rapid but somewhat predictable topology changes, with frequent fragmentation, a small effective network diameter, and redundancy that is limited temporally and functionally.

VANETS are considered as one of the most prominent technologies for improving the efficiency and safety of modern transportation systems. For example, vehicles can communicate detour, traffic accident, and congestion information with nearby vehicles early to reduce traffic jam near the affected areas. VANETs applications enable vehicles to connect to the Internet to obtain real time news, traffic, and weather reports. VANETs also fuel the vast opportunities in online vehicle entertainments such as gaming and file sharing via the Internet or the local ad hoc networks.

Applications such as safety messaging are near-space applications, where vehicles in close proximity, typically of the order of few meters, exchange status information to increase safety awareness. The aim is to enhance safety by alerting of emergency conditions. Applications for VANETs are mainly oriented to safety issues (*e.g.*, traffic services, alarm and warning messaging, audio / video streaming and generalized infotainment, in order to improve the quality of transportation through time-critical safety and traffic management applications, [1]). At the same time, also entertainment applications are increasing (*e.g.*, video streaming and video-on-demand, web browsing and Internet access to passengers to enjoy the trip).

Applications of alarm messaging have strict latency constraints of the order of few milliseconds, and very high reliability requirements. In contrast, applications such as traffic and congestion monitoring require collecting information from vehicles that span multiple kilometers. The latency requirements for data delivery are relatively relaxed *i.e.*, they are "delay-tolerant", however, the physical scope of data exchange is much larger. In contrast, general purpose Internet access requires connectivity to the backbone network via infrastructure, such as Road-Side Units (RSUs).

Non-safety applications are expected to create new commercial opportunities by increasing market penetration of the technology and making it more cost effective. Moreover, comfort and infotainment applications aim to provide road travelers with needed information support and entertainment to make the journey more pleasant. They are so varied and ranges from traditional IP-based applications (*e.g.*, media streaming, voice over IP, web browsing, etc.) to applications unique to the vehicular environment (*e.g.*, point of interest advertisements, maps download, parking payments, automatic tolling services, etc.).

More in general, we can distinguish between *intra* and *inter*-vehicle communications. The first term is used to describe communications within a vehicle, while the second one represents communications between vehicles, or vehicles and sensors, placed in or on various locations, such as roadways, signs, parking areas, and so on. Inter-vehicle communications can be considered to be more technically challenging because vehicle communications need to be supported both when vehicles are stationary and when they are moving. As an instance, the use of a prepaid or automatic billing system when a vehicle slows down instead of stopping at a toll-booth is provided by using a small electronic transmitter. Another example is the integration of cameras and speed sensors to determine the speed of a vehicle (*i.e.*, the well-known autovelox).

Quality of service provided in a VANET is strongly affected by mobility of vehicles, and then dynamic changes of network topology. Different classes of vehicles can move in VANETs, depending on traffic conditions (*i.e.*, dense and sparse traffic), speed limits in particular roads (*i.e.*, highways, rural roads, urban neighborhoods), and also typology of vehicles (*i.e.*, trucks, cars, motorcycles, and bicycles). In general, compared to traditional mobile nodes in MANETs, vehicles in VANETs move at higher speeds (*i.e.*, from 0 to 40 m/s).

All these unique features let VANETs well fit into the class of *opportunistic networks* that means the network behavior is changing and connectivity availability is not always satisfied. As a typical example, in order to maintain network connectivity in VANETs, it is a common technique to connect vehicles traveling on the roadway in opposite directions by means of opportunistic connectivity links. This situation is described as *bridging* technique [3]. However, link breakages strongly hinder stable and durable V2V communications, and as a result communications are dropped. On the other hand, the limited infrastructure coverage, because of sparse fixed access points settling, may cause short-lived and intermittent V2I connectivity. It follows that *interconnectivity* and *seamless connectivity* issues in vehicular ad hoc networks represent a challenge for many researchers. Solutions based on both horizontal and vertical *handover* procedures have been largely investigated in recent works [4], [5], [6].

To summarize, V2V communications have the following advantages: (*i*) to allow short and medium range communications, (*ii*) to present lower deployment costs, (*iii*) to support short messages delivery, and (*iv*) to minimize latency in the communication link. Nevertheless, V2V communications present the following shortcomings that can be solved with the integration with V2I, such as (*i*) frequent topology partitioning due to high mobility, (*ii*) problems in long range communications, (*iii*) problems using traditional routing protocols, and (*iv*) broadcast storm problems [7] in high density scenarios. On the other hand, the strong points of V2I, are the following: (*a*) information dissemination for VANETs, especially using advanced antennas [8], (*b*) VANET / Cellular interoperability [9], and (*c*) WiMAX (Worldwide Interoperability for Microwave Access) penetration in vehicular scenarios [10]. The integration of WiMAX and WiFi technologies seems to be a feasible option for better and cheaper wireless coverage extension in vehicular networks.

In this chapter we will introduce the state-of-the-art of recent technologies used in vehicular networks, specifically for *smart vehicles*, which require novel functionalities such as data communications, accurate positioning, control and decision monitoring.

This chapter is organized as follows. Section 2 describes the concept of *smart vehicles*, and introduces the main components, sensors and capabilities. Section 3 deals with the technologies for VANETs, such as IEEE 802.11*p*, 3GPP, CALM, and Cognitive Radio. Finally, an overview of main applications in VANETs will be provided in Section 4, while conclusions are drawn at the end of this chapter.

2. Smart vehicles

In the next years, vehicles will be equipped with multi interface cards, as well as sensors, both on board and externally. With an increasing number of vehicles equipped with on-board wireless devices (*e.g.*, UMTS, IEEE 802.11*p*, Bluetooth, etc.) and sensors (*e.g.*, radar, ladar, etc.), efficient transport and management applications are focusing on optimizing flows of vehicles by reducing the travel time ad avoiding any traffic congestions. As an instance, the on-board vehicle radar could be used to sense traffic congestions and automatically slow the vehicle. In other accident warning systems, sensors are used to determine that a crash occurred if air bags were deployed; this information is then relayed via V2V or V2I within the vehicular network.

Forgetting traditional vehicles, in the next few years we will drive *smart —intelligent— vehicles*, with a set of novel functionalities (*e.g.*, data communications and sharing, positioning information, sensor equipment, etc.). It is then necessary that for specific applications (*i.e.*, safety messages and alerts, gossip-based applications, etc.) the majority of mobile vehicles within a vehicular network be equipped with *on-board* wireless device, namely On-Board eqUipment (OBU).

A number of systems and sensors are used to provide different levels of functionality. Among the major systems and sensors exploited for intra-vehicle communications we cite: the braking system, crash sensors, the data recorder, the engine control unit, the electronic stability control, the electronic steering, the infotainment system, the integrated starter generator, the lighting system, the power distribution and connectivity, seat belt sensors, the tire pressure monitoring system, etc., [11]. Particularly, for the brake systems, there are also the parking brake and the antilock brake system. The parking brake, which is also referred to as an emergency brake, controls the rear brakes through a series of steel cables. This allows the vehicle to be stopped in the event of a total brake failure. Moreover, also vehicle-mounted cameras are largely used to display images on the vehicle console.

Commonly, a smart vehicle is equipped with the following devices and technologies: (*i*) a Central Processing Unit (CPU) that implements the applications and communication protocols; (*ii*) a wireless transceiver for data transmissions among vehicles (V2V) and from vehicles to RSUs (V2I); (*iii*) a Global Positioning Service (GPS) receiver for positioning and navigation services; (*iv*) different sensors laying inside and outside the vehicle to measure various parameters (*i.e.*, speed, acceleration, distance from neighboring vehicles, etc.); (*v*) an input/output interface for human interaction with the system.

The basic idea of *smart vehicles* is addressed to safety issues, and then by a proper combination of functionalities like *control*, *communications*, and *computing* technologies, it will be possible to assist driver decisions, and also prevent wrong driver's behaviors [12]. The *control* functionality is added directly into smart vehicles to connect the vehicle's electronic equipment. The technology used for control should take into account the need of limit vehicle weight; as a matter, the added wiring increases vehicle weight, and weakens performance. It has been proven that for an average well-tuned vehicle, every extra 50 kilograms of wiring —or extra 100 W of power— increases fuel consumption by 0.2 liters for each 100 kilometers traveled.

Based on such considerations, today *control* and *communications* in a vehicular ad hoc network counter the problems of large amounts of discrete wiring. In the following Figure 1 we show the sheer number of systems and applications contained in a modern vehicle's network architecture.

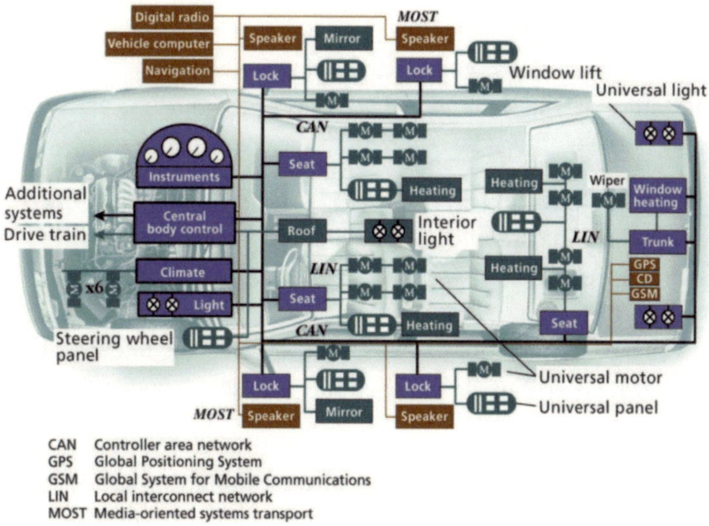

Figure 1. Design of a modern vehicle's network architecture, [13].

In the mid-1980s, Bosch developed the Controller Area Network (CAN), one of the first and most enduring automotive control networks, and now being used in many other industrial automation and control applications. CAN (ISO 11898) is currently the most widely used vehicular network, with more than 100 million CAN nodes sold in 2000.

CAN is a high-integrity serial data communications bus for real-time applications, operating at data rates of up to 1 Mbit/s and having excellent error detection and confinement capabilities. A typical vehicle can contain two or three separate CANs operating at different transmission rates. A low-speed CAN running at less than 125 Kbps usually manages a car's "comfort electronics," like seat and window movement controls and other user interfaces. Generally, control applications that are not real-time critical use this low-speed network segment. Low-speed CANs have an energy-saving sleep mode in which nodes stop their oscillators until a CAN message awakens them. Sleep mode prevents the battery from running down when the ignition is turned off. A higher-speed CAN runs more real-time critical functions such as engine management, antilock brakes, and cruise control. Although capable of a maximum baud rate of 1 Mbps, the electromagnetic radiation on twisted-pair cables that results from a CAN's high-speed operation makes providing electromagnetic shielding in excess of 500 Kb/s too expensive.

CAN is a robust, cost-effective general control network, but certain niche applications demand more specialized control networks. For example, X-by-wire systems use electronics, rather than mechanical or hydraulic means, to control a system. These systems require highly reliable networks.

In 2011 a novel enhanced version of CAN, called *CAN with Flexible Data-Rate* (CAN FD), supports payloads higher than 8 byte per frame. CAN FD protocol controllers are also able to perform standard CAN communication: this allows the use of CAN FD in specific operation modes, like software download at end-of-line programming, while other controllers that do not support CAN FD are kept in standby. In automotive electronics, engine control units, sensors, anti-skid-systems, etc. are connected using CAN. At the same time, CAN is cost effective to build into vehicle body electronics, *e.g.* lamp clusters, electric windows etc. to replace the wiring harness otherwise required.

Another component as shown in Figure 1 is the *LIN-Bus* (Local Interconnect Network), such as a vehicle bus standard –a computer networking bus-system– used within current automotive network architectures. The LIN specification is enforced by the LIN-consortium, with the first exploited version 1.1, released in 1999. The LIN bus is a small and slow network system that is used as a cheap sub-network of a CAN bus to integrate intelligent sensor devices or actuators. Recently LIN is also used over the vehicle's battery power-line with a special DC-LIN transceiver.

A well-known communication system designed for automotive applications is the *FlexRay Communications System i.e.*, a robust, scalable, deterministic, and fault-tolerant digital serial bus system. The core concept of the FlexRay protocol is a time-triggered approach to network communications. This is a different approach to some earlier successful networking schemes. Indeed, FlexRay is an option for upgrading existing network systems using CAN in the automotive industry.

It could be useful for applications, where safety and reliability in a work environment is of most importance due to its deterministic approach and two channel topologies. Due to its high data rate of 10Mb/s over each of its two channels, this protocol suits as the basis of a network backbone. The FlexRay protocol developed by the FlexRay consortium has already found applications in the automotive industry and looks set to become the network scheme favoured especially in x-by-wire applications and other safety critical systems. There is on-going research into the migration from CAN based systems to FlexRay based systems and as such the protocol could find itself being used in many areas outside the automotive industry. With its deterministic time-triggered approach and the high data rates achievable it is also suitable for safety and control applications.

In recent past, more number of multimedia and telematics applications has been integrated into premium class vehicles. These include Sound system, CD player, navigation systems, video players, voice input. High bandwidth requirement of all these applications has led to the development of infotainment communication system, *i.e. MOST* (Media Oriented Systems Transport). MOST is the de-facto standard for *multimedia* and *infotainment* networking in the automotive industry. The technology was designed from the ground up to provide an efficient and cost-effective fabric to transmit audio, video, data and control information between any devices attached even to the harsh environment of an automobile.

MOST technology is the result of the collaboration between car makers and suppliers, working to establish and refine a common standard within the MOST cooperation. MOST Cooperation was founded in partnership by BMW, Daimler Benz, Becker and OASIS silicon system.

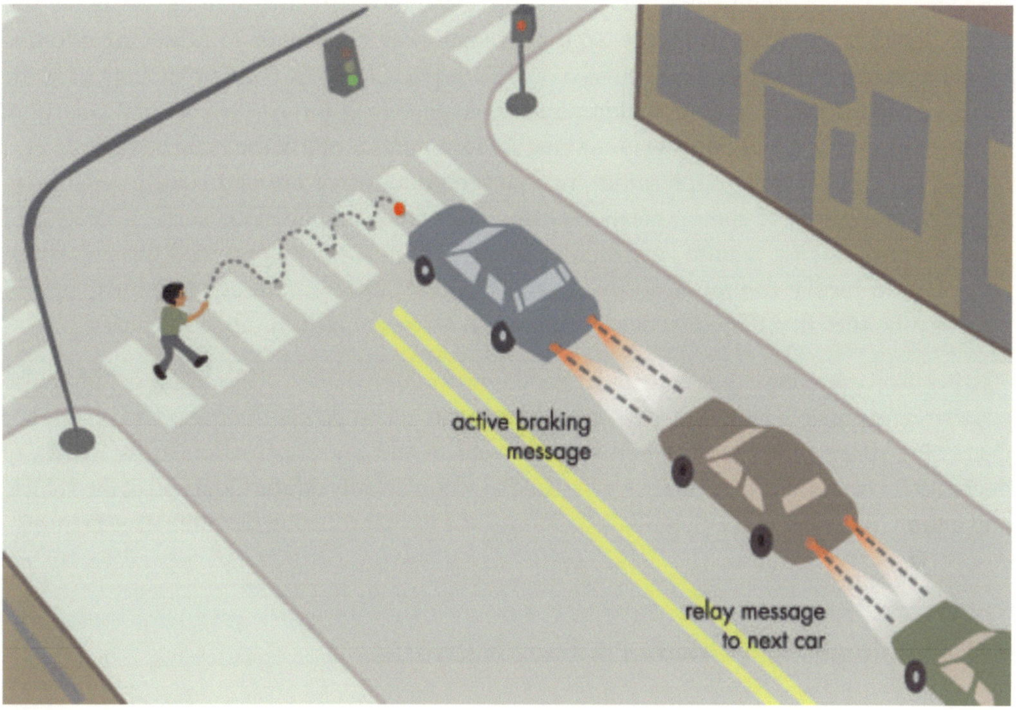

Figure 2. Safety application (*i.e.*, brake messaging) by using VLC devices.

Now SMSC is a leading provider of MOST, the de-facto standard for high bandwidth automotive multimedia networking.

Finally, our attention will be focused on the use of Visible Lighting Communications (VLC) can provide a valid technology for communication purposes in VANETs. The use of the visible spectrum provides service in densities exceeding femtocells for wireless access. It represents a viable alternative that can achieve high data rates, while also providing illumination. This configuration minimizes packet collisions due to Line Of Sight (LOS) property of light and promises to alleviate the wireless bottleneck that exists when there is a high density of rich-media devices seeking to receive data from the wired network.

Possible applications of VLC impact the quality of life, including control of auto / traffic signaling for safety and enabling communications where high noise interferes with WiFi. However, the main issues are related to medium-range LED-based communication derived from outdoor lighting (*i.e.*, sunlight and interference illuminations) that is common in public and private infrastructure (*e.g.*, street, building, and signage illumination).

It is also common on vehicles and in traffic infrastructure including car, rail and air transportation. Specific scenarios include V2V, V2I, as well as V2X communications [14], all of them supporting goals of improved safety, reduced carbon emissions, energy conservation, enhanced connectivity and network performance.

Although LEDs are commonly used in automotive and infrastructure lighting (*i.e.*, brake and traffic lights, as depicted in Figure 2), there remain key challenges to achieving effective modulation and communication between devices, especially while they are moving or in the presence of sunlight. Enabling communications in mobile outdoor systems, particularly in dense, fast moving safety-critical automotive environments is one of the main benefits of VLC for VANETs. In vehicular applications, mobile communications are particularly suitable for adoption of directional communications using LOS links. Applications such as safety and emergency messaging require very high reliability, and this can be provided through short-range inter-vehicular communications. As an instance, vehicles can be equipped with optical transceivers, such that they can communicate with other similarly equipped vehicles.

As a practical example, it is easy to understand how through the use of electronic or light pulse or radar, it is possible to measure the round-trip delay when pulses bounce off an object. For decreasing delays, the distance between the vehicle and any obstruction in the roadway decreases as well. This information can be used to automatically adjust the speed of the vehicle and inform the driver of the potential occurrence of a frontal collision condition (see Figure 2).

3. Technologies in Vehicular Ad hoc Networks

Several technologies are involved in Vehicular Ad hoc Networks, especially as enablers of Intelligent Transportation Systems (ITS). These are GSM, UMTS, Wi-MAX limited Wi-Fi and a new and specific technology thought for this kind of applications, namely Wireless Access in Vehicular Environments (WAVE), also known as IEEE 802.11/*p* [11]. This implicitly suggests that a car should have on board different radio interfaces (and/or network card). About WAVE, it is member of the IEEE 802.11 family, this implicitly suggests that this solution (currently at the stage of draft) is borrowed from IEEE 802.11 and adapted for the vehicular context.

Recent advances in the area of ITS have developed the novel Dedicated Short Range Communication (DSRC) protocol, which is designed to support high speed, low latency V2V, and V2I communications, using the IEEE 802.11*p* and WAVE standards. In 1999, the Federal Communication Commission (FCC) allocated a frequency spectrum for V2BV and V2I wireless communication. DSRC is a communication service that uses the 5.850-5.925 GHz band for the use of public safety and private applications [11].

The allocated frequency and newly developed services enable vehicles and roadside beacons to form VANETs in which the nodes can communicate wirelessly with each other without central access point. Specifically, the communication is in the bandwidth 5.850 GHz—5.925 GHz, so allowing a band of 70 MHz with some guard bands. This band is partitioned in 7 different sub-bands presenting a bandwidth of 10 MHz. The first channel is for V2V communication with public safety purposes while the second and third channels are private channels and used for public safety too in a medium range environment. The fourth is a control channel while the fifth and sixth channels are for public safety services with short range. The seventh channel is dedicated to manage public safety intersections.

Data-rates offered by VANET strictly depend on the kind of service and its own specifications. As an example the smaller data rate is for toll and payment services (highways) where the transmission ate is of the order of few Mb/s at tens of meters. The same distance range is for the Internet access even if the required rate can rise till to 54 Mb/s. Safety message service should allow proactive actions so the range is much higher, of the order of hundreds of meters and the required rate is below 20 Mb/s, down to 6 Mb/s. Finally, services regarding emergency vehicles require rate of the order of 5 Mb/s at a very high distance with respect to previous ones (*e.g.*, 3000 m).

The worldwide ISO TC204 / WG16 has produced a series of draft standards, known as CALM (Continuous Air-Interface, Long and Medium Range). The main goal of CALM is to develop a standardized networking terminal, able to seamless connect vehicles and roadside systems, avoiding disconnections. This can be well accomplished through the use of a wide range of communication devices and networks, such as mobile terminals, wireless local area networks, and the short-range microwave (DSRC) or infrared (IR).

The CALM architecture separates service provision from medium provision via an IPv6 networking layer, with media handover, and will support services using 2G, 3G, 5 GHz, 60 GHz, IEEE 802.16*e*, IEEE 802.20, etc. A standardized set of air interface protocol is provided for the best use of resources available for short, medium and long-range, safety critical communications, using one or more of several media, with multipoint (mesh) transfer.

The CALM concept is now at the core of several major EU sixth framework research and development projects. In the United States, the Vehicle Infrastructure Integration (VII) initiative will be operating using IEEE 802.11*p* / 1609 standards at 5.9 GHz, which are expected to be aligned with CALM 5.9-GHz standards, although the IEEE standards do not have media handover.

Due to the recent strides made in VANETs, a new class of in-car entertainment systems and enabling emergency services using opportunistic spectrum has increased by means of Cognitive Radio (CR) technology [15]. These CR-enabled Vehicles (CRVs) have the ability to use additional spectrum opportunities outside the IEEE 802.11*p* specified standard band.

The growing spectrum-scarcity problem, due to the request of high-bandwidth multimedia applications (*e.g.*, video streaming) for in-car entertainment, and for driver-support services, such as multimedia-enabled assistance, has driven researchers to use the CR technology, for opportunistic spectrum use, which directly benefits various forms of vehicular communication. In such a network, each CRV implements spectrum management functionalities to (*i*) detect spectrum opportunities over digital television frequency bands in the Ultra-High Frequency (UHF) range, (*ii*) decide the channel to use based on the QoS requests of the applications, and (*iii*) transmit on it, but without causing any harmful interference to the licensed owners of the spectrum.

CRVs have many unique characteristics that involve additional considerations than merely placing a CR within a vehicle. As an example, unlike static CR systems, the spectrum availability perceived by each moving vehicle changes dynamically over time, as a function not only of the activities of the licensed or Primary Users (PUs) but also based on the relative motion between them. Spectrum measurements need to be undertaken over the general movement path of the vehicles, leading to a path-specific distribution, instead of focusing on the temporal axis alone.

The CRV network can also leverage the constrained nature of motion, *i.e.* along linear and predecided paths corresponding to streets and freeways. At busy hours or in urban areas, spectrum information can be exchanged over multiple cooperating vehicles, leading to know more about the spectrum availability.

This also allows the vehicles that follow to adapt their operations and undertake a proactive response, which is infeasible in both static and non-stationary scenarios with random motion. The CRV networks fall under three broad classes, such as (*i*) V2V only, (*ii*) V2I only, and (*iii*) centralized V2I. More in detail, in the first class, a network can be formed between vehicles only that rely on cooperation for increasing accuracy. The second class deals with periodic interactions between vehicles and roadside BSs, where the latter acts as a repository of data that is subsequently used by passing vehicles. Finally, a completely centralized network is possible, in which the BS autonomously decides the channels to be used by the CRVs, without relying on information from the vehicles.

The access problem can be solved on two different layers. The first access problem is the selection of the network providing the service. The second one is the access within the selected network. This is mainly true for the V2I environment. Once specified the network quality metric, the vehicle should select the best network (this is the principle of *vertical handover*). Regarding the V2V communications, requiring network synchronization appears complicated so static access procedure x-DMA usage is by fact discouraged. Dynamic access best suit the typical channel features of the multi-hop network so Carrier Sensing Multiple Access – Collision Avoidance may be used. In this context the lack of synchronization at network level is not dramatic and requires only a node-by-node synchronization.

About routing procedures, these are a key point since, especially in the V2V scenario, each kind of message generated after a sensing action should be forwarded, in principle, to all the interested vehicles. In the V2I environment, the routing is not so critical even if vertical handover procedures should be considered.

Regarding V2V connections some routing *philosophies* can be considered. These are Geo-Broadcast, when a node send to all its neighbors an update about a region, Geo-anycast when a vehicle interrogates other nodes about road status and Fleetnet Routing, when a Greedy approach is used, that is, each node tries to forward the information according to a metric (variation of flooding) and it can be implemented via a beacon-based scheme that requires to each node to periodically transmit its position. It this last the positioning is really important and it can be derived or via an absolute service —like GPS— or by triangulation, so requiring more signaling.

The use of GPS (and, more in general, the GNSS) unit within the vehicles allows knowing the vehicles' positions. The awareness of precise locations is very important to every vehicle in VANET so that it can provide accurate data to its neighbors. Currently, typical localization techniques integrate GPS (GNSS) receiver data and measurements of the vehicle's motion.

http://dx.doi.org/10.5772/55492

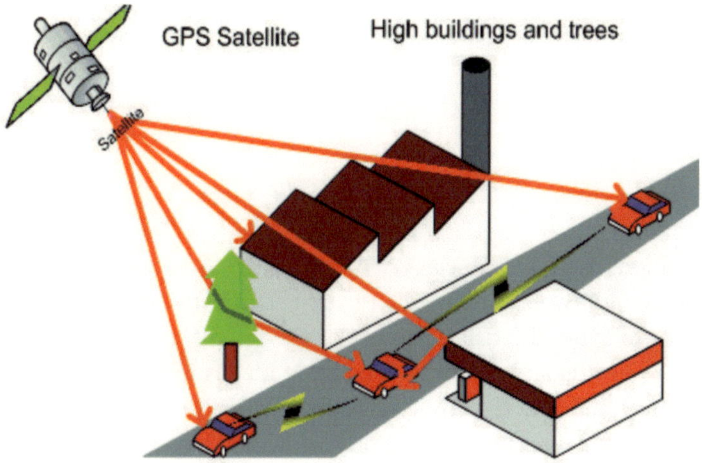

Figure 3. The use of satellite (GPS system) for outdoor localization. However, multipath effect affects the accuracy, [16].

GPS is a positioning system developed and operated by the U.S. Department of Defence. A GPS system is formed from a network of satellites that transmit continuous coded information, which makes it possible to identify locations on Earth by measuring distances from the satellites. At the same time, the receiver has the ability to obtain information about its velocity and direction.

With respect to VANET, many techniques have been proposed to the use of GPS as a localization technique, as shown in Figure 3. However, in many applications a simple GPS receiver is not a satisfactory tool for location estimation (*e.g.*, to discriminate vehicles between those vehicles on a particular highway and others outside the highway), also due to the multipath effect, specially affecting urban areas. Such a system requires highly accurate location estimation. Solutions integrating a GPS with an Inertial Navigation System (INS) can increase the accuracy of the localization application. Also augmented GPS solutions like Differential GPS are largely used for increasing accuracy.

Finally, hybrid approaches comprising of both V2V and V2I mode are largely used in order to improve network performance, while limiting the packet forwarding delay. For example, the ubiquitous integration of existing high-speed WLANs with wide-range 3GPP Long Term Evolution (LTE) results in the service extension of the backbone cellular network, [17]. LTE is the upcoming 4G cellular network with high data rate support for multimedia services, and robustness to high speed. The use of small cells will be massively deployed for increasing coverage areas; as a result they can be good candidates for V2I communications due to their reduced cost.

Novel solutions exploit the use of LTE technology in VANETs. In [18], the authors propose LTE4V2X approach, for the framework of a centralized vehicular network, whose effectiveness has been proven with respect to decentralized protocols. LTE4V2X uses both the IEEE 802.11*p* and 3GPP, LTE to provide an efficient way to periodically collect messages from vehicles and send them to a central server. As a result, the use of heterogeneous wireless network architectures achieves seamless data connectivity among separated vehicular clusters.

4. Applications

Vehicular applications are typically classified in (*i*) active road *safety* applications, (*ii*) traffic efficiency and *management* applications, and (*iii*) *comfort* and *infotainment* applications [11]. The first category aims to avoid the risk of car accidents and make safer driving by distributing information about hazards and obstacles. The basic idea is to broaden the driver's range of perception, allowing him/her to react much quicker, thanks to alerts reception through wireless communications. The second category focus on optimizing flows of vehicles by reducing travel time and avoiding traffic jam situations. Applications like enhanced route guidance/navigation, traffic light optimal scheduling, and lane merging assistance, are intended to optimize routes, while also providing a reduction of gas emissions and fuel consumption.

Finally, although the primary purpose of VANETs is to enable safety applications, non-safety applications are expected to create commercial opportunities by increasing the number of vehicles equipped with *on-board* wireless devices. Comfort and infotainment applications aim to provide the road traveler with information support and entertainment to make the journey more pleasant. In the next subsections we will describe the main aspects of *safety* and *entertainment* applications for VANETs.

The applications regarding *safety* are strictly tied to the main purpose of vehicles: moving from a point till to destination. Car collisions are currently one of the most frequent dead causes and it is expected that till 2020 they will become the third cause. This leads to a great business opportunity for infotainment, traffic advisory service, and car assistance.

Safety applications are always paramount to significantly reduce the number of accidents, the main focus of which is to avoid accidents from happening in the first place. For example, TrafficView [15] and StreetSmart [18] inform drivers through vehicular communications of the traffic conditions in their close proximity and farther down the road. Vehicle platooning is another way to improve road safety. By eliminating the hassle of changing lane and/or adjusting speed, platooning allows vehicles to travel closely yet safely together [20]. Fuel economy can also benefit from reduced aerodynamic as a vehicle headway is tightened (*e.g.*, the spacing can be less than 2 m [21]). Together with adaptive cruise control assisted by V2V communications, the problem of vehicle crashes due to human error can be alleviated.

Some of the most requested applications by polls, currently under investigation by several car manufacturers are Post Crash Notification (PCN), Congestion Road Notification (CRN), Lane Change Assistance (LCA) and Cooperative Collision Warning (CCW). In the following, a brief overview of the above-cited applications is provided.

In PCN, a vehicle involved in an accident would broadcast warning messages about its position to trailing vehicles so that it can take decision with time in hand as well as to the highway patrol for asking away support. The PCN application may be implemented both on V2V and V2I network configurations. In fact the V2V presents the advantage of giving quickly the information through a *discover-and-share* policy. Through the use of specific sensors, it consists in measuring possible changes in the rational behavior of the driver (*e.g.*, quick brake use, rapid direction changes, and so on), which are then communicated back via directional antennas to the other vehicles along the same direction. Once received, the closest vehicle can share this information with the other nodes with a flooding routing. In the particular case of false alarm by the first vehicle experiencing the irrational behavior of the driver, this information floods on the VANET. It is then important to fix the issue of false alarms.

Let us suppose a driver has been distracted by something on the panorama and moves the steering wheel, so that the vehicle direction changes accidentally. Once recognized the error, the driver will react by quickly changing direction or with a quick and strong use of breaks. This behavior is not rational since there is no danger for the VANET community, but only the behavior of a single is irrational. This represents a false indication of alarm. If the first following driver does not experience some accidents, then the vehicle does not forward this information, and false alarm probability is reduced, otherwise if it discovers the same problem, it shares such information with the other vehicles.

Dealing with the use of V2I architecture, the access points should gather information (*e.g.*, alarms for quick speed changes), coming from different vehicles, and merging the data so reducing the signaling from the vehicles. The V2V has the drawback of not allowing a quick communication if the vehicles are far away from each other (*e.g.*, in low traffic density scenarios), while the V2I is more energy consuming since it should be on all the time.

The LCA application constantly monitors the area behind the car when passing or changing lanes, and warns the driver about vehicles approaching from the rear or in the next lane over. This application has two different modalities, the first one is the so called passive mode, while the other one is the active mode. In the passive mode the vehicle simply measures distances, by means of detection and ranging procedures, while in the active mode it communicates to the other vehicles that they are too close, so they should change their direction / behavior.

Traffic monitoring and management are essential to maximize road capacity and avoid traffic congestion. Crossing intersections in city streets can be tricky and dangerous at times. Traffic light scheduling can facilitate drivers to cross intersections. Allowing a smooth flow of traffic can greatly increase vehicle throughput and reduce travel time [22]. A token-based intersection traffic management scheme is presented in [23], in which each vehicle waits for a token before entering an intersection. On the other hand, with knowledge of traffic conditions, drivers can optimize their driving routes, whereby the problem of (highway) traffic congestion can be lessened [24].

CRN detects and notifies about road congestions, which can be used for route and trip planning. This kind of application is partially implemented in current GPS-based applications where a new route is evaluated when heavy congestion has been detected on a route or in a portion of it. Up till now several commercial tools are available for smart-phones and special purpose devices. These are currently based on GPS coordinates and local resident software able to indicate the shortest or fastest routes from a starting point till to a destination by considering one ways streets and so forth. A new generation of this kind of software integrates some control messages coming from the so-called Radio Data System-Traffic Message Channel (RDS-TMC) that gathers information about unavailable routes or congested streets. TMC messages contain a considerable amount of information:

- *Identification*: what is causing the traffic problem and its seriousness;
- *Location*: the area, road or specific location affected;
- *Direction*: the traffic directions affected;
- *Extent*: how far the problem stretches back in each direction;
- *Duration*: how long the problem is expected to affect traffic flow;
- *Diversion advice*: alternative routes to avoid the congestion.

The service provider encodes the message and sends it to FM radio broadcasters, who transmit it as an RDS (Radio Data System) signal within normal FM radio transmissions. There's usually only about 30 seconds between the first report of an incident to the traffic information centre and the RDS-TMC receiver getting the message.

Also this application may be implemented according to a V2V configuration or a V2I one. In fact, it is possible to encapsulate information about the position, the direction, and the average speed, which are then communicated back to the vehicle following on the street the information. As it appears clear, this solution suffers for a large amount of data to be processed by the vehicles themselves. What is worth in this environment is the use of V2I since the access points can process information coming and communicate to the incoming vehicles the new route after request information about their destination. So, with a software that implements what is current available on the market (with a special purpose processor in this case and without strict bounds on energy consumption for processing) it is possible to develop an instance of ITS.

Finally, the CCW system works with a cutout revealing a stopped car, or a stopped or slow-moving car before arrival at the curve or downhill. All these applications require radio transceivers for message exchange, GPS and sensor on board car and road infrastructure units. Even in this case the dualism between V2V and V2I is renovated. Not so different from PCN, the behavior of the driver must be *understood* by the vehicles and then forwarded to the following cars and the vehicles coming with an opposite direction. This can be set up in a V2V modality since once recognized (*i.e.*, just happened) it is important to spread and flood this information. In the second phase, about 30 seconds or one minute, depending on traffic level, the information should be managed by the access point so to advice upcoming vehicles in time.

For what concerns *non-safety* applications, they have very different communication requirements, from no special real-time requirements of traveller information support applications, to guaranteed Quality-of-Service needs for multimedia and interactive entertainment applications. In general, this class of applications is motivated by the desire of the passengers to communicate either with other vehicles or with ground-based destinations (*e.g.*, Internet hosts or the Public Service Telephone Network (PSTN)). Also, various traveler information applications belong to this category. As an instance, the driver could receive local information regarding restaurants, hotels and, in general, Point of Interest, whenever the vehicle is approaching there (*i.e., context aware* applications).

The aim of *infotainment* applications is to offer convenience and comfort to drivers and/or passengers. For example, Fleetnet [25] provides a platform for peer-to-peer file transfer and gaming on the road.

A real-time parking navigation system is proposed in [26] to inform drivers of any available parking space. Digital billboards for vehicular networks are proposed in [27] for advertisement. Internet access can be provided through V2I communications; therefore, business activities can be performed as usual in a vehicular environment, realizing the notion of mobile office [7]. On-the-road media streaming between vehicles also can be available [7], [8], making long travel more pleasant. An envisioned goal is to embed *human-vehicle-interfaces,* such as color reconfigurable head-up and head-down displays, and large touch screen active matrix Liquid Crystal Displays (LCDs), for high-quality video-streaming services. Passengers can enjoy their traveling time by means of real-time applications *e.g.,* video streaming and online gaming, using individual terminals next to their seats [28]. Figure 4 (a) and (b) depict the use of LCD devices for entertainment applications.

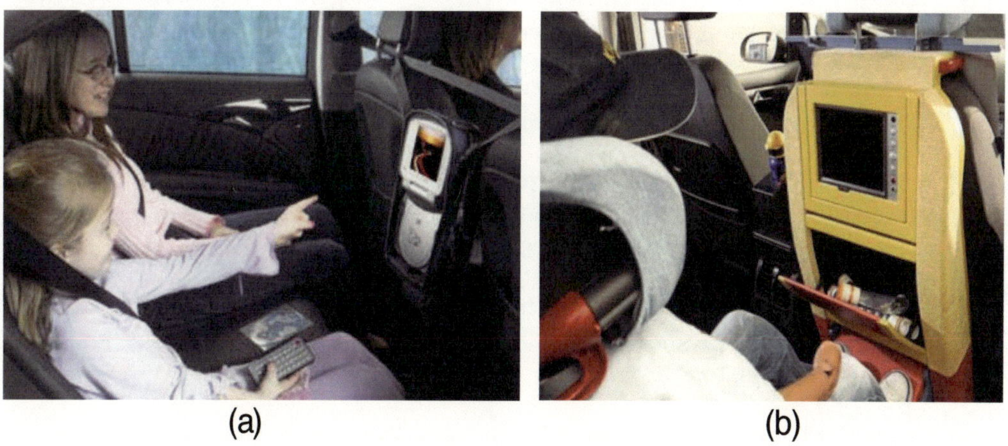

Figure 4. Video-streaming applications for passengers in a smart vehicle, [28].

5. Conclusions

In this chapter we have presented the main technologies used in vehicular networks for *smart vehicles*. Different technologies can be exploited to provide communication, control and safety capability to vehicles, as well as comfort and entertainment applications are well supported.

We investigated main aspects of vehicular ad hoc networks (*i.e.,* the communication protocols V2V and V2I, differences from MANETs, main applications), and the main technologies, and sensors, used to support emerging inter and intra-vehicle communications. We also envisaged the vehicle of tomorrow *i.e., smart* vehicle, which will be considerably different from the vehicle of today, due to the use of short distance transmitters and receivers that operate similarly to Doppler radar, and the increase in a vehicle's ability to control and take decisions.

Author details

Anna Maria Vegni[1], Mauro Biagi[2] and Roberto Cusani[2]

1 University of Roma Tre, Department of Applied Electronics, Rome, Italy

2 University of Rome Sapienza, Department of Information Engineering, Electronics and Telecommunications, Rome, Italy

References

[1] VANET Vehicular Applications and Inter-Networking TechnologiesEd. H. Hartein-stein, K.P. Labertaux), John Wiley & Sons, Ltd., Mar. (2010).

[2] Lind, R. et al., The Network Vehicle-a glimpse into the future of mobile multi-media, IEEE Aerospace and Electronic Systems Magazine, Sept. (1999). , 14(9), 27-32.

[3] Agarwal, A, & Little, T. D. C. Opportunistic Networking in Delay Tolerant Vehicular Ad Hoc Networks, In Advances in Vehicular Ad-Hoc Networks: Developments and Challenges (Ed. M. Watfa), (2010). , 282-300.

[4] Vegni, A. M, & Cusani, R. Connectivity Support in Heterogeneous Wireless Networks, in Recent Advances in Wireless Communications and Networks, Edited by Jia-Chin Lin, 978-9-53307-274-6INTECH Pub, August (2011).

[5] Vegni, A. M, Inzerilli, T, & Cusani, R. Seamless Connectivity Techniques in Vehicular Ad-hoc Networks, in Advances in Vehicular Networking Technologies, Edited by Miguel Almeida, 978-9-53307-241-8INTECH Pub, April (2011).

[6] Inzerilli, T, Vegni, A. M, Neri, A, Cusani, R, & Cross-layer, A. Location-Based Approach for Mobile-Controlled Connectivity, Int. Journal of Digital Multimedia Broadcasting, Hindawi Publ. Corp., Article ID 597105, 13 pages, doi:(2010). , 2010

[7] Soldo, F, Casetti, C, Chiasserini, C. F, & Chaparro, P. Streaming Media Distribution in VANETs, In Proc. of the IEEE GLOBECOM, November-December (2008). , 1-6.

[8] Tseng, Y, Ni, C, Chen, S. -Y, & Sheu, Y. -S. J.-P., The Broadcast Storm Problem in a Mobile ad hoc Network. Wireless Networks, 8: (2002). , 153-167.

[9] Kaul, S, Ramachandran, K, Shankar, P, Oh, S, Gruteser, M, Seskar, I, & Nadeem, T. Effect of Antenna Placement and Diversity on Vehicular Network Communications, In 4th Annual IEEE Communications Society Conference on Sensor, Mesh and Ad Hoc Communications and Networks, 2007. SECON'07, (2007)., 112-121.

[10] Sengupta, R, Rezaei, S, Shladover, S. E, Cody, D, Dickey, S, & Krishnan, H. Cross-layer-based Adaptive Vertical Handoff with Predictive RSS in Heterogeneous Wireless Networks, Vehicular Technology, IEEE Transactions on, 7(6): (2008)., 3679-3692.

[11] Moustafa, H, & Zhang, Y. Vehicular Networks: Techniques, Standards, and Applications, Auerbach Publications, Taylor and Francis Group, 450 pages, Ch. 2, (2009).

[12] Varaiya, P. Smart cars on smart roads: problems of control," IEEE Transactions on Automatic Control, Feb (1993)., 38(2), 195-207.

[13] Leen, G, & Heffernan, D. Expanding Automotive Electronic Systems, Computer, Jan. (2002)., 35(1), 88-93.

[14] Vegni, A. M, & Little, T. D. C. Hybrid Vehicular Communications Based on V2V-V2I Protocol Switching, Int. Journal of Vehicle Information and Communication Systems (IJVICS), Nos. 3/4, , 2, 213-231.

[15] Di Felice MDoost-Mohammady R., Chowdhury K.R., Bononi L. Smart Radios for Smart Vehicles: Cognitive Vehicular Networks, IEEE Vehicular Technology Magazine, June (2012)., 7(2), 26-33.

[16] Drawil, N. Improving the VANET Vehicles' Localization Accuracy Using GPS Receiver in Multipath Environments, Master of Applied Science, available online http://libdspace.uwaterloo.ca/bitstream/10012/3349/1/Nabil_thesis_MASc_UW.pdf

[17] Sivaraj, R, Gopalakrishna, A. K, Chandra, M. G, & Balamuralidhar, P. QoS-enabled group communication in integrated VANET-LTE heterogeneous wireless networks, Wireless and Mobile Computing, Networking and Communications (WiMob), 2011 IEEE 7th International Conference on, vol., no., Oct. (2011)., 17-24.

[18] Remy, G, Senouci, S, Jan, F, Gourhant, Y, & Lte, V. X: LTE for a Centralized VANET Organization, Global Telecommunications Conference (GLOBECOM 2011), 2011 IEEE, vol., no., Dec. (2011)., 1-6.

[19] Nadeem, T, Dashtinezhad, S, Liao, C, & Iftode, L. TrafficView: traffic data dissemination using car-to-car communication, ACM SIGMOBILE Mobile Comput. Commun. Rev. 8 (3) ((2004)., 6-19.

[20] Dornbush, S, & Joshi, A. Street Smart Traffic: Discovering and Disseminating Automobile Congestion using VANETs, In Proc. of the IEEE VTC, Spring, April (2007)., 11-15.

[21] Gehring, O, & Fritz, H. Practical Results of a Longitudinal Control Concept for Truck Platooning with Vehicle-to-Vehicle Communication, In Proc. of the IEEE ITSC, November (1997). , 117-122.

[22] California Partners for Advanced Transit and Highways (PATH) [Online]Available: / http://www.path.berkeley.edu/S.

[23] Huang, Q, & Miller, R. The Design of Reliable Protocols for Wireless Traffic Signal Systems, Technical Report WUCS-02-45, (2008). , 2898-2906.

[24] Dresner, K, & Stone, P. A Multiagent Approach to Autonomous Intersection Management, J. Artif. Intell. Res. 31 (1) ((2008).

[25] Hartenstein, H, Bochow, B, Lott, M, Ebner, A, Radimirsch, M, & Vollmer, D. Position-aware ad hoc Wireless Networks for Inter-vehicle Communications: the Fleetnet project," in: Proceedings of the ACM MobiHoc, (2001). , 259-262.

[26] Verizon Wireless [Online]Available: /http://www.verizonwireless.com/S.

[27] Guo, M, Ammar, M. H, Zegura, E. W. V, & Vehicle-to, a. Vehicle Live Video Streaming Architecture, Pervasive Mobile Comput. 1 (4) ((2005).

[28] "Backseat child navigation concept for kids"available online at http://www.slippery-brick.com.

Section 2

Smart Roads Communication Infrastructures

Chapter 2

Base Station Design and Siting Based on Stochastic Geometry

Hui Zhang, Yifeng Xie, Liang Feng and Ying Fang

Additional information is available at the end of the chapter

http://dx.doi.org/10.5772/55493

1. Introduction

In this chapter, the base station (BS) design and siting method is introduced, which includs three parts: general BS design and siting method, stochastic geometry in BS design and siting, frequency planning for BS design and siting. Moreover, the concept of stochastic geometry is introduced, and also stochastic geometry theory is taken in wireless network analysis.

Stochastic geometry is a helpful method in modeling vehicular ad-hoc network, especially in vehicle to infrastructure communication. In this chapter some related concepts of stochastic geometry are introduced, which may help to establish the model of vehicular ad hoc network, such as vehicle to vehicle communication network.

Vehicular ad hoc networks (VANETs) are a special case of mobile ad hoc networks (MANETs), where such network is formed between vehicles [1]. We can analyze the stochastic model of MANETs to derive the model for VANETs. MANETs are wireless networks made of one type of nodes (Vehicle to Vehicle communication of VANETs is made of vehicles). Each node can either transmit or receive information using the same frequency band. Each node is denoted as the terminal (such as source and destination) or router (such as relay). Moreover, each hop by hop transmission and the connectivity between source and destination depend on the location of intermediate relay nodes [2]. With different distribution of the above nodes, various hot spots can be established. For example, assume these nodes are Poisson point process, written as P.P.P. Then we can use the properties of P.P.P. to derive formulas, such as SINR. [1,3] give some models by means of analyzing Aloha in VANETs using stochastic geometry theory. In actual scenarios, there are many ways to establish different stochastic geometry models for VANETs. It is not practical to deploy a high-density network for vehicle to vehicle (V2V) and vehicle to infrastructure (V2I) communication instantly. New workforce skills for the installation and maintenance of V2I applications need to be considered, as well as the privacy policy restrictions. V2I communication works well if cars are in low speed or keeping relative rest. But with high speed, cars cannot keep information interaction with infrastructures for a long time. Mobile relay station (MRS) is one scheme which can be used in high-speed conditions. MRS provides the information interaction between base station and mobile terminals in cars, reducing the connections from a single mobile terminal in the car to the base station. Group handover scheme and two-level resource allocation algorithm can also be considered in the MRS scheme. Heterogeneous vehicular wireless architecture based on different technologies,such as WAVE (IEEE 802.11p) and WiMAX (IEEE 802.16e) technology, can be a great help for V2I communication.

On the other hand, it's a critical method for frequency planning in BS design and siting. Under this background, the frequency planning theoretical analysis is given in this chapter for OFDMA-based cellular systems, including four parts, respectively multigraph theory, algebraic analysis principle, extension theory and Stackelberg theory. In graph theory, the coloring theory in multi-graph and the level interference-limited theory are focused, resulting in an frequency reuse analysis from the optimization problem. In algebraic analysis principle, a quantitative analytic algebra to describe the relationship of frequency reuse factor between cell center and cell edge is given and the frequency reuse optimization problem is transformed into two-dimensional coordinate system, enables to take analytic algebra method to solve it. In extension theory analysis, the multi-dimensional cell-edge element model and its multi-element extension set is established for frequency allocation. In Stackelberg theory, the frequency optimization in cooperative communication is formulated into Stackelberg problem and a Stackelberg model is established for this architecture. On the basis of the above theoretical analysis, a soft fractional frequency reuse (SFFR) scheme is presented, including two parts: SFFR I and SFFRII. The numerical results show that 个修改定 when FRF is large, while taken SFFRII scheme when FRF is small. Furthermore, it needs to take into account the size of FRF, both cell-edge and cell-center performance to choose the appropriate inner radius.

2. General BS design and siting method

Base station planning is very important in the whole process of wireless network optimization, including base station siting, the configuration of base station equipment, wireless network parameters setting and the analysis of wireless network performance.

2.1. Principle of BS design and siting

In BS design and siting, firstly it needs to determine the number of required BS according to the analysis of link budget and network capacity. The principle of base station design and siting is given as follows:

1. Estimate the amount of base stations by means of link budget and coverage requirement.

http://dx.doi.org/10.5772/55493

2. Analyze the capacity of actual network and then determine the required amount of base stations which can meet the need of capacity.

3. Compare and choose base station according to the results of estimation and analysis.

The base station planning is complex. Both technical and feasible factors should be considered. The configuration of equipment is designed both hardware and software of base stations, based on the coverage, capacity, quality requirement and ability of design. By surveying the expected location of base stations and simulating the whole wireless network, the right wireless parameters can be chosen to meet the requirement of design, including transmission type, antenna height, antenna angle, carrier frequency, etc. Before actual base station design and siting, it's always needed to simulate the performance of the whole wireless network by Monte Carlo method.

In BS planning, the preliminary setting equipment load and carrier types in different area is determined according to prophase demand analysis, which divides into two parts, respectively link budget and capacity planning. There are two formulas, written as N1 and N2. N1 denotes as the number of base stations meeting coverage requirement and N2 denotes as the number of base stations meeting capacity requirement. Comparing N1 with N2, the number of base stations in this area is chosen as N=max (N1, N2). Then the base station planning is done for the total area. On the one hand, if N1<<N2, check whether the equipment overload. If it is, we can increase the carrier frequency of the large capacity demand area. On the other hand, the load can be decreased or increased according to different situations. The process of BS planning is shown in Fig.1.

2.2. Estimate the amount of BS

The basic process of estimating the amount of BS is given as follows:

1. Determine the bearing capacity of the wireless network: derive the bearing capacity of the whole link by simulating and testing the network.

2. Determine the parameters of the whole wireless network.

3. Derive the minimum amount of BS to meet the requirement of coverage. The coverage area of each BS: **R** is the maximum cover radius of each BS. **S** is the coverage area of each BS. **D** is the average distance between two neighboring BSs. Based on antenna properties, BS can be divided into two types: Omni-directional and directional. Many directional BSs are three sectors. The coverage areas are different with different BS types.

4. Derive the minimum amount of BS to meet the requirement of capacity. Table 1 shows the traffic capacity of different BS types.

5. Determine the configuration of BS according to the capacity of each sector.

6. Modify the number of BS for different areas.

Besides, some other factors can be considered in estimating the amount of BS, such as link budget, pre-planning and antenna types.

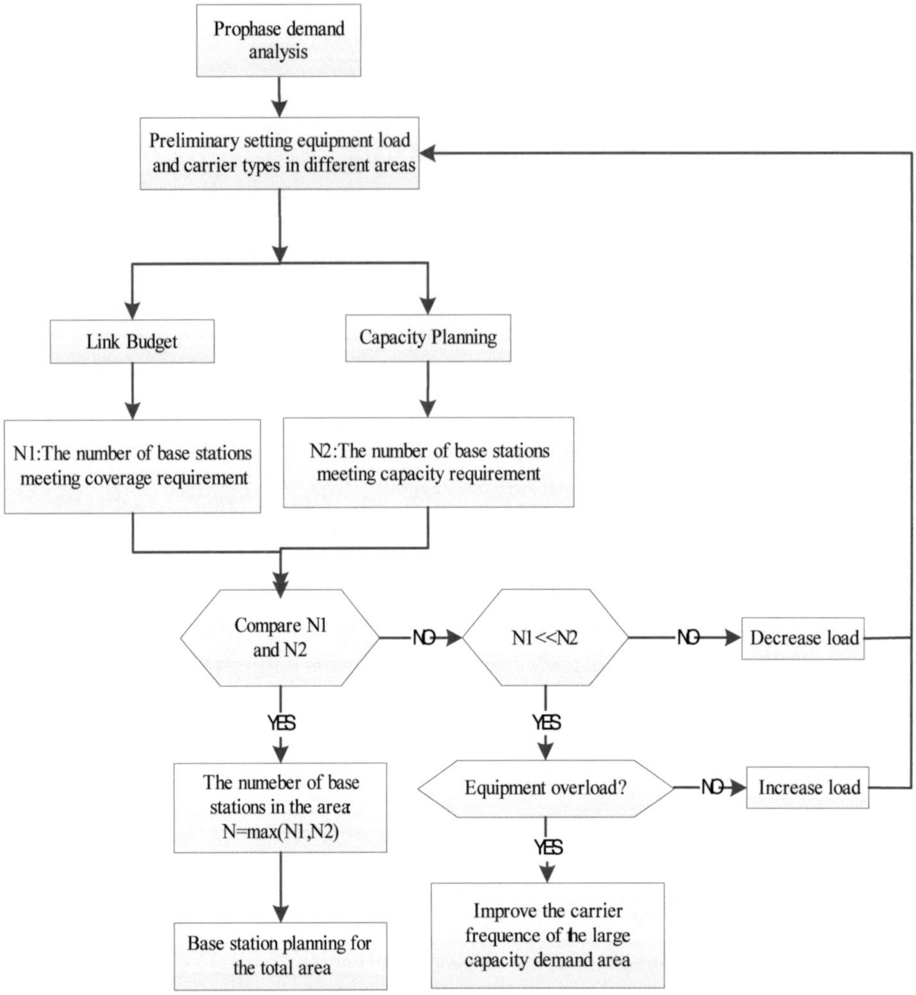

Figure 1. The general process of BS planning

Typs of BSs	Number of Sectors	Number of carrier frequency	Voice Channel(Erl)	Visual telephone(Erl)	Throughput of downlink(kbit/s)
Omni-directional	1	1	DUi	NUi	SVi
directional	3	1	3*DUi	3* NUi	3* SVi
directional	3	2	6*DUi	6* NUi	6* SVi
directional	3	3	9*DUi	9* NUi	9* SVi

Ps : DUi, NUi,SVi (respectively denote as Dense Urban, Nomal Urban, Suburb and Village) are the capacity of single sector of a BS.

Table 1. Traffic capacity of different BS types

i. Omni-directional BS:

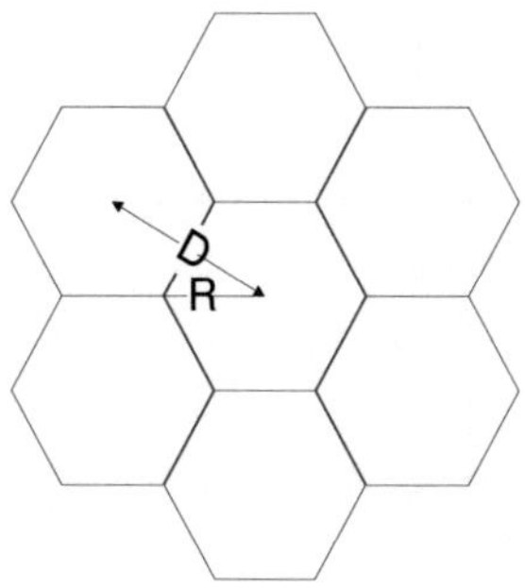

Figure 2. Honeycomb structure of omni-directional BS($S = 3\sqrt{3}R^2 / 2$)

ii. Three sector directional BS(clover)

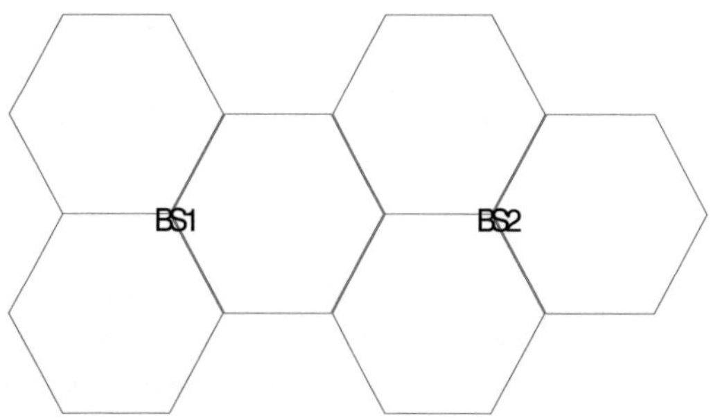

Figure 3. Honeycomb structure of three sector directional BS (clover) ($S = 9\sqrt{3}R^2 / 8$)

This structure is generally used in the downtown.

iii. Three sector directional BS(hexagon)

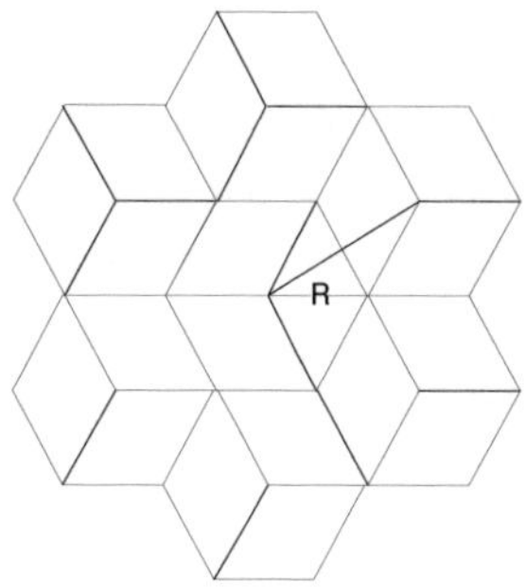

Figure 4. Honeycomb structure of three sector directional BS (hexagon) ($S = 3\sqrt{3}R^2 / 2$)

This structure is generally used in villages.

3. Stochastic geometry in BS design and siting

With the development of communication technology, the density of users is increasing, as well as the interference. As both interference and density of users can be modeled by the spatial location of nodes, mathematical techniques have been used in this area, such as stochastic geometry, including point process theory [4]. In [5], a method is introduced to add optimal number of base stations to the optimal placement of an existing wireless data network, where the whole network is modeled as a P.P.P.

Cellular networks are usually modeled by placing the base stations according to a regular geometry such as a grid, with the mobile users scattered around the network either as a Poisson point process (i.e. uniform distribution) or deterministically. These models have been used extensively for cellular design and analysis but suffer from being both highly idealized and not very tractable. Mathematical analysis for conventional (1-tier) cellular networks is known to be hard, and so highly simplified system models or complex system level simulations are generally used for analysis and design respectively. To make matters worse, cellular networks are becoming increasingly complex due to the deployment of multiple classes of BSs that have distinctly traits.

Cellular networks are in a major transition from a carefully planned set of large tower-mounted base stations to an irregular deployment of heterogeneous infrastructure elements that often additionally includes micro, pico, and femtocells, as well as distributed antennas. For example, a typical 3G or 4G cellular network already has traditional BSs that are long-range and guarantee near-universal coverage, operator-management picocells and distributed antennas that have a more compact form factor, smaller coverage area and are used to increase capacity while eliminating coverage dead zones and femtocells, which have emerged more recently and are distinguished by their end-user installation arbitrary locations, very short range, and possibility of having a dozed-subscriber group. This evolution toward heterogeneity will continue to accelerate due to crushing demands for mobile data traffic caused by the proliferation of data-hungry devices and applications.

In this section, we plan to develop a tractable, flexible and accurate model for a downlink heterogeneous cellular network (HCN) consisting of K tiers of randomly located BSs, where each tier may differ in terms of randomly located BSs, where each tier may differ in terms of average transmit power, supported data rate and BS density.

3.1. Stochastic geometry theory

3.1.1. Poisson point process

A spatial point process (p.p.) Φ is a random, finite or countably-infinite collection of points in the d-dimensional Euclidean space R^d, without accumulation points[6].

Definition: Let Λbe a locally finite non-null measure on R^d. The possion print process $\Phi$of intensity measure $\Lambda$is defined by means of its finite-dimensional distributions:

$$P\{\Phi(A_1)=n_1,...,\Phi(A_k)=n_k\}=\prod_{i=1}^{k}\left(e^{-\Lambda(A_i)}\frac{\Lambda(A_i)^{n_i}}{n_i!}\right) \tag{1}$$

for every k =1,2,... and all bounded, mutually disjoint sets A_i for i = 1,2,...k. If $\Lambda(dx)=\lambda dx$ is a multiple of Lebesgue measure (volume) in R^d, we call Φ is a homogeneous Poisson p.p. and λ is its intensity parameter[6].

Poisson point process is a tractable tool to model the random distribution of users or base stations with high density. Other point processes can also be used to analyze the distribution of base stations with high density.

3.1.2. Voronoi tessellation

In mathematics, a Voronoi tessellation is a special kind of division of a given space, determined by distances to a specified family of objects (subsets) in the space. These objects are usually called the sites or the generators and to each such object one associate a corresponding Voronoi cell, namely the set of all points in the given space whose distance to the given object is not greater than their distance to the other objects.

Definition: Given a simple point measure μ on R^d and a point $x\in R^d$,the Voronoi cell $C_x(\mu)=C_x$ of the point $x\in R^d$ w.r.t μ is defined to be the set

$$C_x(\mu)=\left\{y\in R^d:|y-x|<\inf_{x_i\in\mu,x_i\neq x}|y-x_i|\right\} \tag{2}$$

The Voronoi cell is often defined as the closure of the last set. Given a simple point process $\Phi=\sum_i \varepsilon_{x_i}$ on R^d, the Voronoi Tessellation (VT) generated by Φ is defined to be the marked point process[6].

$$\nu=\sum_i \varepsilon_{\left(x_i,C_{x_i}(\Phi)-x_i\right)} \tag{3}$$

Fig.5 depicts the Voronoi tesselation. The points in Voronoi cell are nearer to the corresponding object, which generates the Voronoi cell, rather than the other objects.

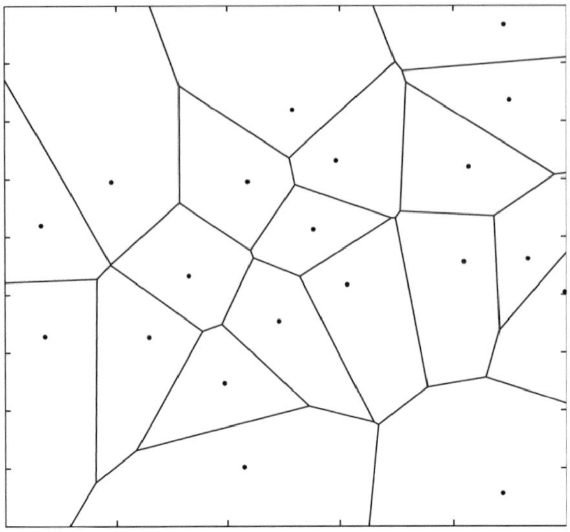

Figure 5. Sample of Voronoi tessellation

A weighted Voronoi diagram is that we assign distinct weight to generator points of normal Voronoi Tessellations and it is more applicable under some situations. A Voronoi region V_i is the intersection of the dominance regions of p_i over every other generator point in P. While we will have different formulae for dominance regions in weighted Voronoi diagrams, the idea remains the same. The dominance region of a generator point p_i over another, p_j, where $i \neq j$ and $d_w(p_x, p_y)$is the weighted distance between points x and y, is written as

$$Dom_w\left(p_i, p_j\right) = \left\{p \mid d_w\left(p, p_i\right) \leq d_w\left(p, p_j\right)\right\} \tag{4}$$

Let

$$V_w\left(p_i\right) = \bigcap_{p_j \in P \setminus \{p_i\}} Dom_w\left(p_i, p_j\right) \tag{5}$$

$V_w(p_i)$ is called as a weighted Voronoi region and $V_w = \{V_w(p_1), V_w(p_2), \ldots, V_w(p_n)\}$ is called as the weighted Voronoi diagram [7].

3.2. Stochastic modeling

The cells served by base stations in mobile communications are far from the regular hexagonal shape which is often taken as a reference model. Then Voronoi tessellation and other knowledge of stochastic geometry will be helpful in analysis.

3.2.1. Basic cell model

In this model, assume the network is just a single hierarchical level, and all subscribers are served by the closest BS, where the Voronoi tessellation method is taken. Considering the configuration of the subscribers, the distribution of BS usually obeys stochastic point process. For the simplicity of analysis, such processes can be considered as Point possion processes.

Take two independent Poisson processes Π_0 and Π_1 as an example, Π_0 and $\Pi_1$represent the subscribers and the stations respectively. The parameters of this model come to be the intensity measures Λ_0 and Λ_1 of two processes. As the density of subscribers is bigger than the one of base stations, we can assume $\Lambda_0=\alpha\Lambda_1$ for some$\alpha>>1$. Fig.6. depicts the basic cell model, where the subscribers are connected with the base station of their corresponding Voronoi cell. The link between subscriber and base station is not shown in the picture.

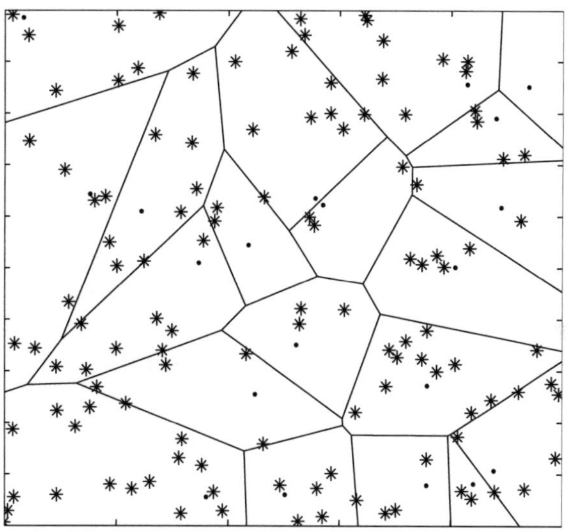

Figure 6. Basic cell model with a single hierarchical level of stations point: base station; *point: subscribers; blue line: cell boundary

3.2.2. Hierarchical model

In this model there are several tiers of stations, and the processes of these tiers are thought to be independent with the decreasing intensity $\lambda_0>\lambda_1>\cdots>\lambda_N$. We can assume stations of tier i are represented by a realization of a homogeneous Poisson process Π_i [8-9].

A real wireless network is more complex.But we can separate it from several tiers, such as macro BSs, pico BSs and femto BSs. BSs are independent from each tier.In this model, there's no connection between each tier. Fig.7. depicts a three tiers model. In this three tiers model, BSs of different tiers are not connected.

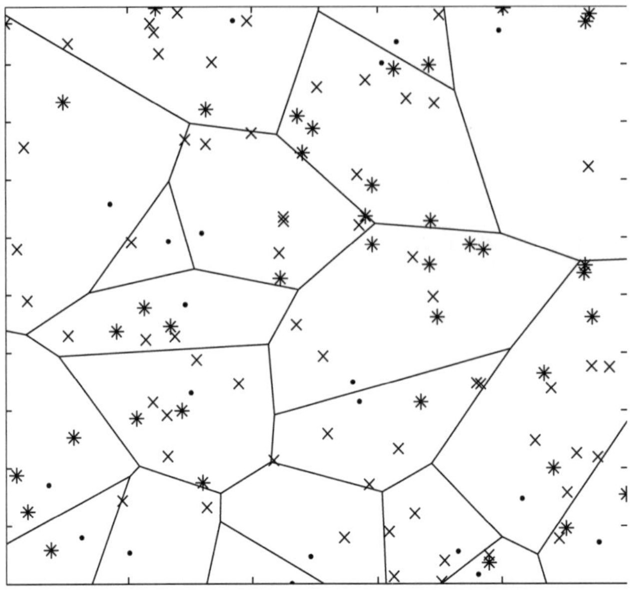

Figure 7. Three tiers mode Macro BSs: points; femto BSs: x points; pico BSs: * points; line: macrocell boundary

Classical methods of communication theory are generally insufficient to analyze new types of networks, such as ad-hoc and sensor works. Stochastic geometry allows us to study the average behavior over many spatial realizations of a network, whose nodes are placed according to some probability distribution. So, it is a good tool for the analysis of new type networks. As the model with stochastic geometry is constructed, it needs to analyze the network performance combined with actual scenarios, which enable to get the useful result we want. More details about the stochastic models and the use of stochastic geometry knowledge to derive theorems in these models will be researched in the future.

4. Frequency planning for BS design and siting

In LTE and its evolution system, many frequency planning schemes are derived from Soft Frequency Reuse (SFR) and Fractional Frequency Reuse (FFR). Kim et al[10]. propose an incremental frequency reuse (IFR) scheme that reuses effectively the radio spectrum through systematic segment allocation over a cluster of adjoining cells, which divides the entire frequency spectrum into several spectrum segments. Li et al[11]. give a cooperative frequency reuse scheme for coordinated multi-point transmission. Liang et al[12]. propose a frequency reuse scheme for OFDMA based two-hop relay enhanced cellular networks. Chen et al[13]. give an approach based on large-scale optimization to deal with networks with irregular cell layout. Wamser et al[14]. give some different strategies for user and resource allocation are evaluated along with fractional frequency reuse schemes in the uplink. Assaad et al[15]. give an analysis of the inter cell interference coordination problem and study the optimal fractional frequency reuse (FFR). Imran et al[16]. propose a novel self-organizing framework for adaptive frequency reuse and deployment in future cellular networks, which forms an optimization problem from spectral efficiency, fairness and energy efficiency. Novlan et al[17]. give a comparion of fractional frequency reuse approaches in the OFDMA cellular downlink, which mainly focuses on evaluating the two main types of FFR deployments, respectively Strict FFR and Soft Frequency Reuse (SFR).

However, the existing research mainly major in the form of application under different cellular scenarios, but lack of analyzing the theoretical basis for each frequency reuse scheme. It's important to find some necessary theories to guide the design of frequency reuse scheme. Considering this problem, this paper makes an analysis of frequency reuse theoretical basis from four angles, respectively multigraph theory, algebraic analysis principle, extension theory and Stackelberg theory. By means of the above theoretical tools, we try to summarize the rules of frequency reuse design, give a way to find the optimal frequency reuse scheme. Furthermore, a soft fractional frequency reuse (SFFR) scheme is introduced, including two parts: SFFR I and SFFR®.

The rest of this part is organized as follows: The multigraph theory, algebraic analysis principle, extension theory and Stackelberg theory are respectively introduced in Section 4.1, Section 4.2, Section 4.3 and Section 4.4. The soft fractional frequency reuse scheme is described in Section 4.5. Then the numerical results are analyzed in Section 4.6.

4.1. Multigraph theory

In [18], the coloring method in graph theory and the collection idea are taken into frequency reuse sets, dividing cellular users into different sets and providing a unique frequency reuse strategy to each set, which efficiently raise cellular frequency reuse factor. Ref.[19] adopts graph theory to discuss the problem of frequency allocation optimization, proposes a k-level interference-limited theory for the whole frequency sets. Moreover, it establishes an optimization model for frequency allocation. Ref.[20] analyzes the cellular frequency planning by multigraph T-coloring method. In graph theory, the concept of multigraph means that every pair of points is at most connected with K edges, also without no self-loop, such graph is written as K-multigraph. The concept of T-coloring can be defined as: In graph G, color each point in the set V(G) by T classes of colors, and the colors are different among each adjacent point, written as T-coloring in G graph. Ref.[21] describes some basic characteristics of frequency allocation in T-coloring. In view of graph theory in frequency optimization, we further analyze the frequency reuse optimization problem using multigraph coloring theory and k-level interference-limited theory.

Assume each cellular cell cluster is divided into n limited districts $\{a_1, a_2, \cdots, a_n\}$, and allocate the frequency $f(a_i)$ into each district a_i. For the reuse of co-frequency resources, the level of interference is always different among districts due to the path loss of inter-cell interference. As shown in Fig.8., the co-frequency interference strength of No. 0 cell to its adjacent cell is inversely proportional to the path distance. By multigraph theory, the interference level can be mapped into interval according to the strength, which enables to optimize the frequency allocation.

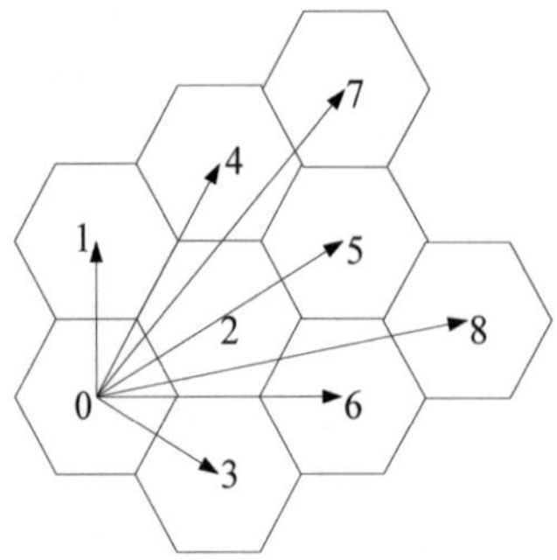

Figure 8. Inter-cell interference level

As shown in Fig.9., in order to analyze the inter-cell co-frequency interference, the interference strength in different districts is divided into several intervals by the descending range $\{[i_1, i_2], [i_2, i_3], \cdots, [i_{m-1}, i_m]\}$, which map into each interference level $l:\{l_1, l_2, \cdots, l_m\}$.

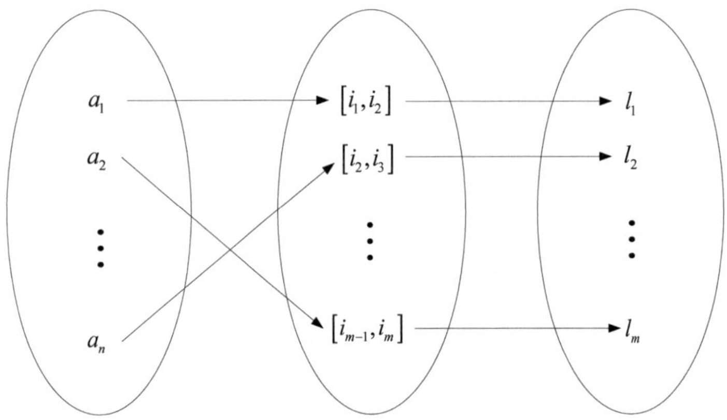

Figure 9. Inter-cell interference level mapping

When District a_u and District a_v exist the interference with the same level, the frequency allocation in this area should satisfy:

$$\{a_u, a_v\} \in E \Rightarrow |f(a_u) - f(a_v)| \notin T(l) \tag{6}$$

In particular, when K = 2, $T(l_0) = \{0\}$, which means the interference among a_u and a_v are the level l_0. In order to optimize frequency allocation, it's necessary to allocate different frequency among a_u and a_v. When K=2, $T(l_1) = \{0, 1\}$, it means the interference among a_u and a_v are in the level l_1, so the frequency allocation for such two area should not only satisfy difference, but also not adjacent.

Furthermore, we consider K different cells $\{G_0, G_1, \cdots, G_{K-1}\}$, and the number of frequency allocation area is V(G) for each cell. Moreover, the co-frequency interference among a_u and a_v are with the same level l. For inter-cell interference level l, we define the taboo collection $T(l)$ for frequency allocation, as follows:

$$G_0 \supseteq G_1 \supseteq \cdots \supseteq G_{K-1} \tag{7}$$

$$T(0) \subseteq T(1) \subseteq \cdots \subseteq T(K-1) \tag{8}$$

In the view of graph theory, the frequency reuse aims to allocate frequency into each point in multigraph, find the allocation function f to make the co-frequency interference minimum. Moreover, it enables to color the entire cell $\{G_0, G_1, \cdots, G_{K-1}\}$ by $T(l)$ color. In other words, the formula (4-1) is established under the function f.

4.2. Principles of algebraic analysis

Based on the multigraph theory, we propose an algebraic analysis method for frequency reuse, which change the relationship of cellular frequency reuse factor into quantitative algebra analytic formula, taking two-dimensional coordinates to solve this frequency reuse optimization problem.

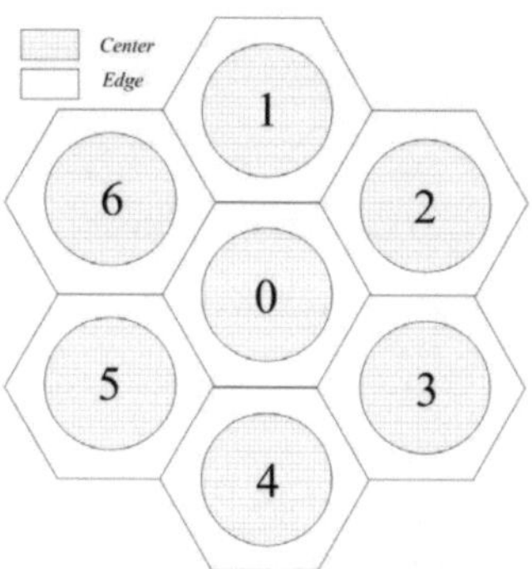

Figure 10. Cellular frequency reuse

As shown in Fig.10., considering soft frequency reuse (SFR), taking No.0 Cell as an example, we define the cell-center region 0C (0 Center, simply written as 0C) as variable y, the cell-edge region 0E (0 Edge, simply written as 0E) as variable x. In this way, we take two-dimensional coordinates to analyze the function of x and y. Since 0C is in the cell-center region, its maximum value of frequency reuse factor (FRF) is equivalent to 1, so the variable y should satisfy:

$$0 < y \leq 1 \tag{9}$$

0E is in the cell-edge region and its FRF is related with relevant partition way for cell-edge. When its FRF takes 1/k (k=3, 7… n), the variable should satisfy:

$$0 < x \le \frac{1}{k} \tag{10}$$

For 0C and 0E are still in the same cell, the total sum of FRF for the whole cell should be not more than 1, that is

$$0 < x + y \le 1 \tag{11}$$

At the same time, when there's not divided for cell-center and cell-edge, the total FRF can be expressed as

$$0 < x + y \le \frac{1}{k} \tag{12}$$

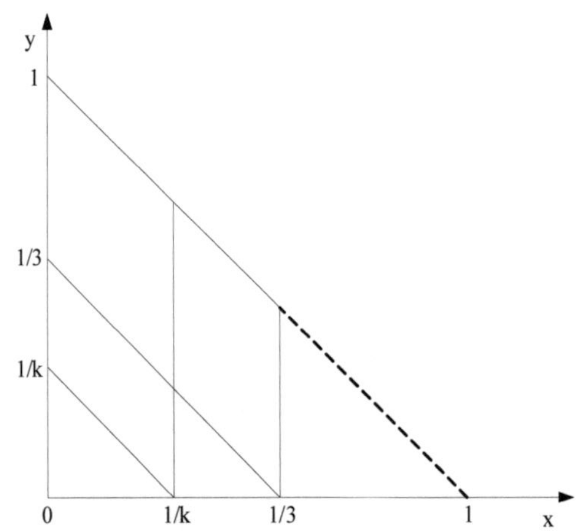

Figure 11. Frequency reuse frequency reuse factor coordinates

Based on the above analysis, as shown in Fig.11., we take the sub-function as follows:

$$\begin{cases} 0 < x + y \le 1, 0 < x \le \frac{1}{k}, 0 < y \le 1 \\ 0 < x + y \le \frac{1}{k}, 0 < x \le \frac{1}{k}, 0 < y \le \frac{1}{k} \end{cases} \tag{13}$$

The Algebraic analysis approach opens a new way to the analysis of cellular frequency reuse and optimization. Based on this idea, 0C and its reuse region {1E, 2E, 3E, 4E, 5E, 6E}, 0E and its interference region {1E, 2E, 3E, 4E, 5E, 6E} can be described into a form of two-dimensional coordinates, which enables to theoretically search the optimal cellular frequency reuse scheme.

4.3. Extension theory in frequency allocation

The frequency allocation in cellular systems could be modeled as the extension set in extension theory [22], by means of which we establish multi-dimensional cell-edge element model. Assume N_i^j denotes as the part of cellular area, written as$(i,\ j)$. c_m denotes as the available probability for frequency set $\bar{f}_m$ in N_i^j, written as $c_m(N_i^j) \in \{\pm 1,\ 0\}$. On the other hand, $c_m(N_i^j) = \pm 1$respectively means whether f_m is taken in N_i^j, while $c_m(N_i^j)=0$ means the critical state. According to the analysis of the inter-cell interference situation in cell-edge, $c_m(N_i^j)$ could be defined as:

1. Initialization$c_m(N_i^j)=0$
2. If $m=j$, then $c_m(N_i^j)=1$, and assign the sub-frequency setf_j into N_i^j
3. According to $c_m(N_i^j) \times c_m(N_j^i) = -1$, determine the other $c_m(N_i^j)$ values, making the co-frequency set not be reused in the adjacent cell-edge area, enabling to reduce the inter-cell interference.

On this basis, it can be obtained:

$$c_m(N_i^j) = \begin{cases} 1, & m = j \\ -1, & m = i \\ 0, & others \end{cases} \tag{14}$$

Therefore, the n-dimension element model is established for the cell-edge sub-frequency set in N_i^j, that is:

$$R_i^j = \begin{bmatrix} N_i^j, & c_1, & c_1(N_i^j) \\ & c_2, & c_2(N_i^j) \\ & \cdots & \cdots \\ & c_n, & c_n(N_i^j) \end{bmatrix} \tag{15}$$

Define the correlation function in the extention set as:

$$K(c_m(R_i)) = \sum_{j=1}^{n} c_m(N_i^j),\ i \neq j \tag{16}$$

where $\sum_{j=1}^{n} c_m(N_i^j)$ denotes as the number of cell-edge area blocks. $K(c_m(R_i))>0$ means f_m avariable in Cell i, $K(c_m(R_i))<0$ means f_m unavariable in Cell i and $K(c_m(R_i))=0$ means the critical state for f_m. Based on the above analysis, the extention function is defined as:

$$T_k K(c_m(R_i)) = K(c_m(R_i)) \times \left(- \sum_{j=1, j \neq i}^{n} c_m(N_j^i) \right) \tag{17}$$

Where $\sum_{j=1, j \neq i}^{n} c_m(N_j^i)$ represents the cell block area adjacent to the edge of the frequency of collection can be used in the area blocks. By means of extension theory, the frequency allocation in cellular system could be modeled by element model and multi-element extension set, enabling to frequency allocation optimization.

4.4. Stackelberg theory in frequency allocation

Recently, cellular cooperative communication for multiple base stations and multiple users is drawing attention as a solution to achieve high system throughput in cell-edge, such as cooperative beam, cooperative resource control, cooperative transmission, cooperative relaying, etc. As shown in Fig.12., it gives the network topology of cellular cooperative communication system, where several access points (AP) are connected into eNodeB and some cell-edge users are served by the cooperative AP. Moreover, as discussed in standardizing groups of IMT-advanced, cooperative communication technologies are expected to be essential in the next generation cellular networks. In this section, we consider taking Stackelberg theory to solve frequency allocation problem in OFDMA-based cooperative cellular systems.

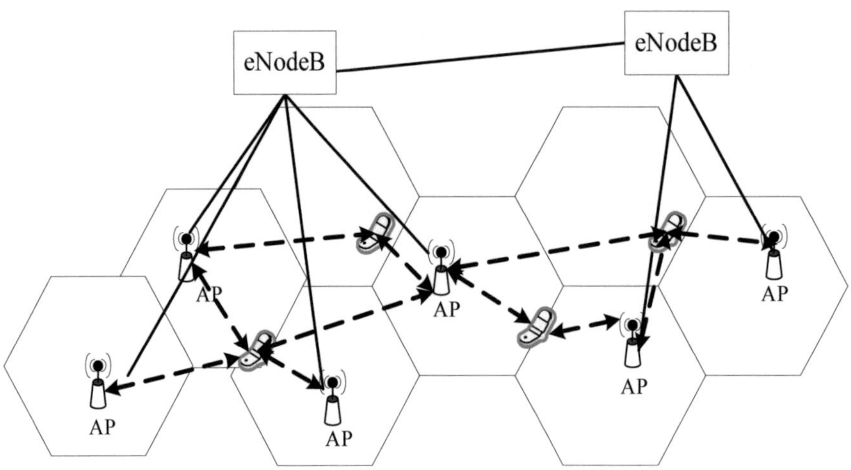

Figure 12. Cellular cooperative communication system

In 1952, Stackelberg [23] presented master-slave hierarchical decision-making issues, also known as Stackelberg problem. Its basic features include: Several independent decision-makers in decision-making, and each decision makers own some controlled decision variables. Some decisions by decision-maker may affect with each other. The decision-making system is always with a hierarchical structure and many decision-makers distribute at different decision-making levels (Fig.13.). The decision could be adjusted from upper decision-maker to lower decision-maker according to self-decision objective. Also, the decision by lower decision-maker affects upper decision-maker. Moreover, master-slave relationship exists among upper/ lower decision-maker, and the optimum decision should satisfy both upper and lower decision-maker.

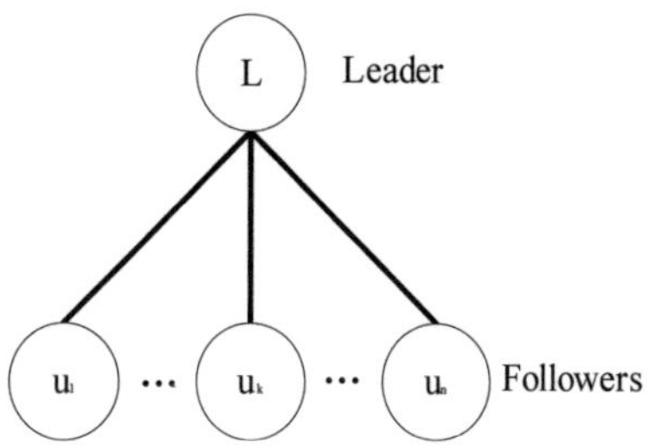

Figure 13. Stackelberg problem

Master-slave model with single-objective decision-making

$$\begin{aligned} & \underset{x}{Max}\ U_0(x,y) \\ & \quad \text{s.t. } (x,y) \in \Omega_0 \\ & \quad\quad \underset{y}{Max}\ U_1(x,y) \\ & \quad\quad\quad \text{s.t. } (x,y) \in \Omega_1 \end{aligned} \tag{18}$$

where x, Ω_0, $U_0(x, y)$ are the main decision variables, constraint set and utility function for the master. y, Ω_1, $U_1(x, y)$ are the decision variables, constraints set and utility function for the slave. Usually, Stackelberg problem is with no-convex properties, including two categories: general two decision-making algorithm and two linear optimization algorithm. For the former, the algorithms based on penalty function and KKT conditions can be taken. For the latter, the algorithms based on KKT conditions and linear programming optimization can be taken.

Furthermore, we summarize the mathematical model of cooperation communication into Stackelberg problem: the primary unit L is decision-maker, the cooperative unit F_i is policy maker, and the number of cooperative nodes is N_L , meeting $i=1, ..., N_L$. The uplink of UE j and F_i is written as i. the utility is written as U_{ij}. By solving the maximum utility value, it can effectively solve the uplink power control problem based on Stackelberg equilibrium for cooperative communication, especially uplink power control problem based on Stackelberg equilibrium for cooperative communication.

4.5. Soft fractional frequency reuse

As shown in Fig.14., the characteristics of soft fractional frequency reuse (SFFR) scheme I is given as follows: the whole cell is divided into two parts, cell-center and cell-edge. In cell-center, the frequency reuse factor (FRF) is set as 1, while in cell-edge FRF is dynamic and the frequency allocation is orthogonal with the edge of other cells, which can avoid some inter-cell interference in cell-edge. On the other hand, users in every cell are divided into two major groups according to their geometry factor: cell-edge users who are interference-limited and caused by the neighboring cells, and cell-center users who are noise-limited. The available frequency resources in cell-edge are divided into some non-crossing subsets in SFFR I.

Since the cell-edge users are easily subject to serious interference, the frequency assignments to the cell-edge users greatly rely on radio link performance and system throughputs. Generally, the cell-edge can be split into 12 parts marked by 1, 4 and 9, just as the cell-edge of Cell 1 in Fig.7. For three adjacent cells, there are 9 parts in the cell-edge corner, which are in the shaded area. Moreover, we take this SFFR I model as an example to deduce the design of the available frequency band assignment for the fields marked by 1,2,..,9. In cell-edge, select frequency from the subsets u_1, u_2 , u_3. If it's not enough, add frequency from u_4, u_5 , u_6. If inter-cell interference becomes serious, increase frequency in u_4, u_5, u_6, and decrease the cover area in cell-edge. When the interference is controlled in a low extension, decrease the frequency in subsets of u_4, u_5, u_6, and increase the cover area in cell-edge, enabling to improve the frequency utilization.

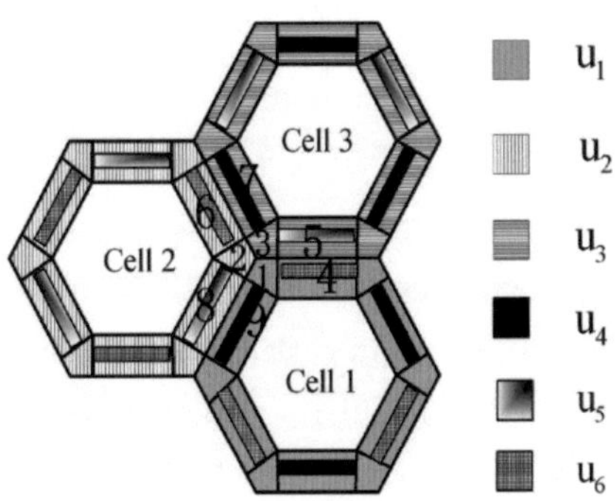

Figure 14. Example of SFFR I scheme

As shown in Fig.15., the users in cell of SFFR II scheme are also divided in to two groups, respectively cell-center users and cell-edge users. Moreover, the available sub-carriers are divided into two no-overlapping parts: G and F. Besides, G is the available sub-carriers for cell-center users. Moreover, the FRF in cell-center is 1. F is the available sub-carriers for cell-edge uers, which is divided into 3 orthogonal parts, respectively u_1, u_2, u_3, enabling to avoid inter-cell interference.

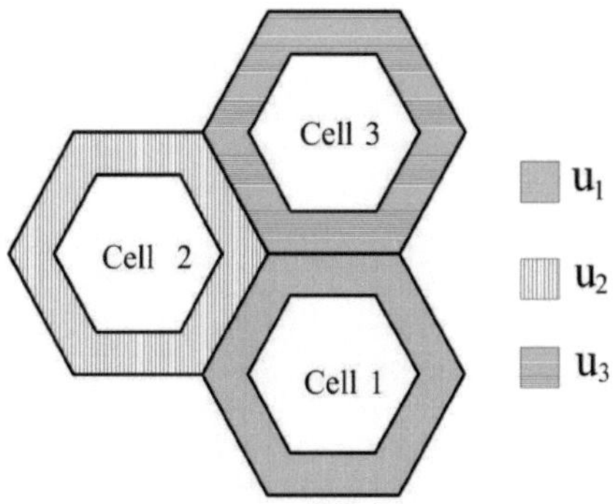

Figure 15. Example of SFFR II scheme

4.6. Performance analysis

4.6.1. Performance analysis with increased users for SFFR

Fig.16. shows the system throughputs as the number of users increases, respectively for SFFR I and SFFR II when FRF is equal to 2/3. Fig.17. shows the average rates for cell-edge and cell-center users. Moreover, the ratio of inner radius and cell radius is 0.8. Some analysis results about Fig.16. and Fig.17. are given as follows:

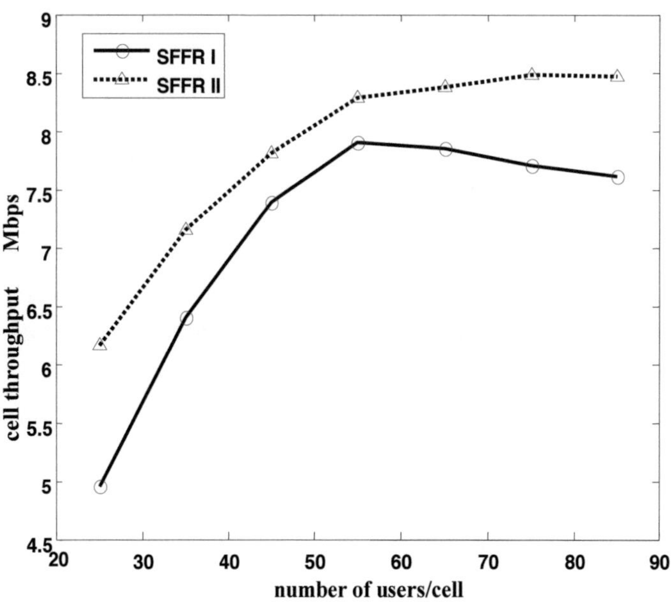

Figure 16. The system total throughputs(FRF = 2/3)

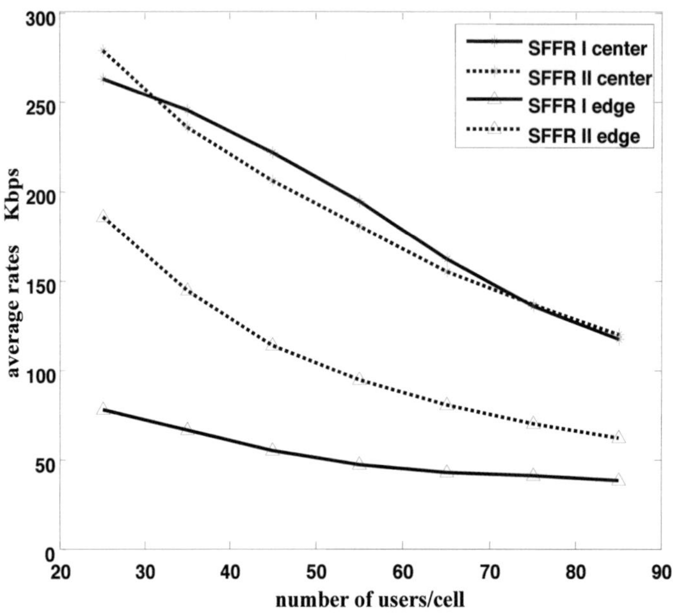

Figure 17. The average rates in cell-center and cell-edge (FRF = 2/3)

1. As shown in Fig.16, the system throughputs increase with the number of users increasing for such two SFFR schemes, but the gradient of the above curves decreases. The reason is that the probability of co-frequency intercell interference may increase as the users increase. As the sub-carrier resources are taken up, the throughputs become saturated or even decreased. On the other side, the blocking probability also increases for there's no available sub-carriers allocated to the new users, which makes the actual number of users may not be increased.

2. As shown in Fig.17, the cell-edge and cell-center average rates gradually decrease as the users increase. The reason is that the inter-cell interference increases, making the C/I decrease and users' performance degrade. Similarly, partial users may be blocked for there's not enough sub-carriers be allocated, making the average rates decrease.

3. With the comparison of such two schemes, for FRF is equal to 2/3, SFFR II is better than SFFR I in the performance of total throughputs and cell-edge average rates, while the cell-center average rates is similar for SFFR II and SFFR I. In the SFFR I scheme, there's overlap for the frequency allocation in cell-edge, while the different frequency sub-carriers are allocated into cell-edge users for the SFFR II scheme. Under this background, the co-frequency interference for cell-edge users is worse in SFFR I scheme than SFFR II scheme.

In addition, the cell-center users in the SFFR I scheme could take use of the available sub-carriers in cell-edge, reducing the blocking probability. At the same time, it makes cell-edge users be interfered more from the adjacent cell-center users. Because of the difference of frequency resource assignments, the performance in cell-edge is better in SFFRII than in SFFR I. On the other hand, it's flexible for the SFFRII scheme, which could decrease the blocking probability in cell-center users for lack of sub-carriers. When the load is small, the cell-center Performance may be a little better in SFFRII than in SFFR I, but as the number of users increases, the performance of such two schemes gradually reach to similar level. In a whole, the total throughputs in SFFRII are more than in SFFR I.

4.6.2. Performance analysis with increased FRFs for SFFR

The total throughputs, cell-edge average rates and cell-center average rates for SFFR I and SFFRII are respectively given in Fig.18., Fig.19. and Fig.20., where the ratio of inner radius and cell radius is 0.8 and the number of users per cell is 65.

It can been seen from Fig.18, Fig.19 and Fig.20 that as the increase of FRF, both the total throughputs and the cell-center average rates increase, while the cell-edge average rates decrease. One reason is that for cell-edge users, the number of available sub-carriers in cell-edge decreases with the FRF increasing, which makes partial cell-edge users not be allocated to enough sub-carriers and be blocked further. On the other hand, more sub-carriers could be allocated to cell-center users, reducing the inter-cell interference and improving the average rates. Moreover, the average rates in cell-center are higher than in cell-edge. On the contrary, with the FRF decreases, the available sub-carriers in cell-edge increase, while decrease in cell-center. As a result, the performance is improved for cell-edge users, while degraded for cell-center users. In order to balance the spectrum efficiency and cell-edge performance, it's necessary to set an appropriate FRF value for cellular system.

When the FRF is in a small value, the available sub-carriers decrease in cell-center, which make the cell-center performance of SFFR I scheme show better than SFFRII scheme. However, the cell-edge performance of SFFR I shows worse than SFFRII for more sub-carriers are reused in cell-center and the cell-edge users in SFFR I are interfered more seriously. On the contrary, When the FRF is in a large value, the available sub-carriers increase in cell-edge, which make the cell-edge performance of SFFR I scheme show better than SFFRII scheme, but the cell-center performance of SFFR I shows worse than SFFRII. According to the analysis, it can be taken SFFR I scheme when FRF is large, while taken SFFRII scheme when FRF is small.

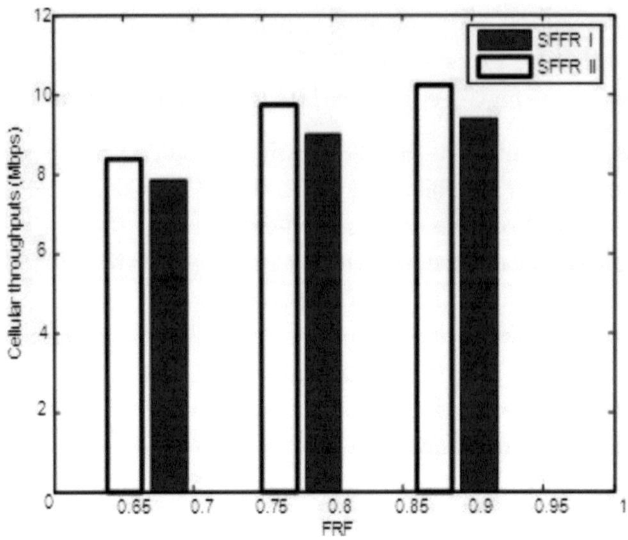

Figure 18. Cellular throughputs of different FRFs (2/3, 7/9, 8/9)

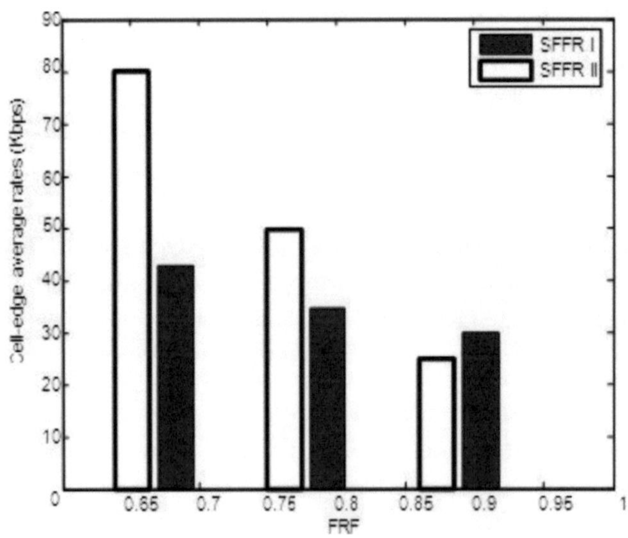

Figure 19. Cell-edge average rates of different FRFs (2/3, 7/9, 8/9)

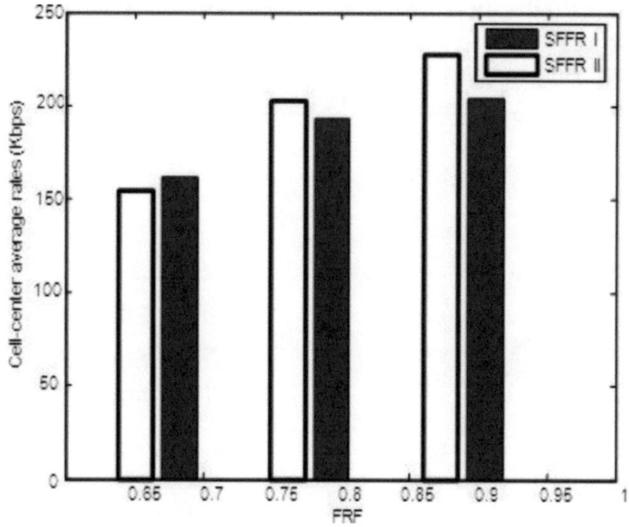

Figure 20. Cell-center average rates of different FRFs (2/3, 7/9, 8/9)

4.6.3. Performance analysis of SFFR

As shown in Fig.21, with the inner radius increasing, the number of cell-edge user decreases, making its average rate increasing, improving the performance of cell-edge user. The larger the FRF is, the better the performance is.

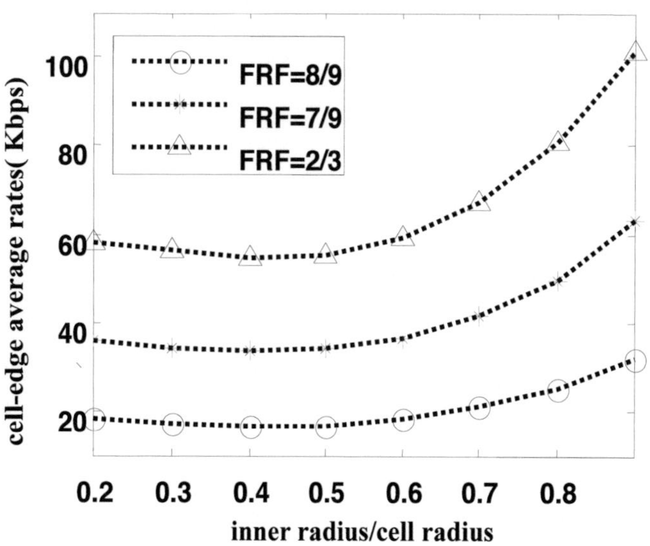

Figure 21. The average rate for cell-edge user (SFR)

As shown in Fig.22., with the inner radius increasing, more users are classified as cell-center users. For each fixed FRF, the frequency resources in cell-center is relatively constant, the increase of cell-center users may lead to insufficient resource allocation and larger co-frequency interference, so the average rate for cell-center user will decrease as the inner radius increases.

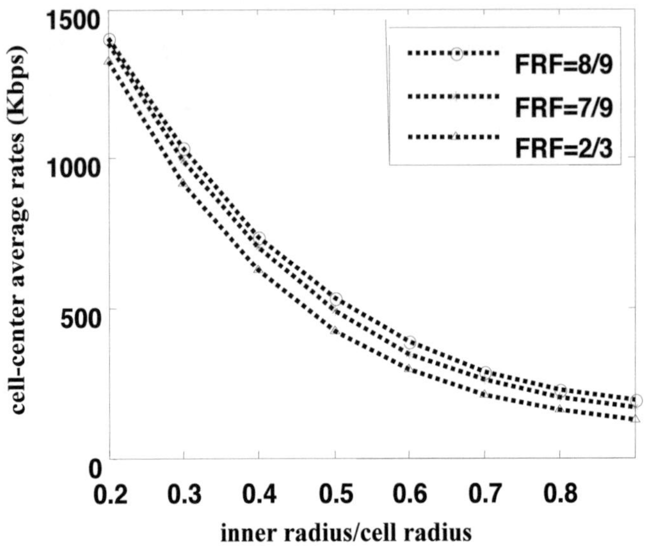

Figure 22. The average rate for cell-center user (SFR)

Therefore, it needs to take into account the size of FRF, both cell-edge and cell-center performance to choose the appropriate inner radius. With the FRF increasing, the inner radius should be increased, making a rational division of cell-center users and cell-edge users.

5. Conclusions

In this chapter, the general BS design and siting method is introduced, which gives the general process of BS planning and the estimation method of BS number. Then, the basic theory about stochastic geometry is introduced in BS design and siting, respectively Poisson point process and Voronoi tessellation. Moreover, the stochastic modeling principle is also given for BS design and siting, respectively basic cell model and hierarchical model. Besides, this chapter gives an analysis on the frequency planning scheme in BS design and siting, and tries to find the theoretical basis of frequency planning from four parts, respectively multigraph theory, algebraic analysis principle, extension theory and Stackelberg theory.

In graph theory, we focus on coloring theory in multigraph and the level interference-limited theory, making an analysis of optimization problem in frequency reuse. In algebraic analysis principle, this proposal gives a quantitative analytic algebra to describe the relationship of frequency reuse factor between cell center and cell edge, and the frequency reuse optimization problem can be transformed into two-dimensional coordinate system, which enables to consider analytic algebra method to solve it. In extension theory analysis, the multi-dimensional cell-edge element model is established and the results show that frequency allocation in cellular system could be modeled by element model and its multi-element extension set. In Stackelberg theory, we formulate the frequency optimization in cooperative communication into Stackelberg problem and establish a Stackelberg model for this architecture.

On the basis of the above theoretical analysis, a soft fractional frequency reuse (SFFR) scheme is presented, including two parts: SFFR I and SFFRII. Moreover, the simulation is taken to compare SFFR performance with different FRF schemes, and the results show that it can be taken SFFR I scheme when FRF is large, while taken SFFRII scheme when FRF is small. Furthermore, it needs to take into account the size of FRF, both cell-edge and cell-center performance to choose the appropriate inner radius. In the future, we would further take Monte Carlo system simulation to compare different frequency reuse schemes and find the optimal scheme according to the proposed method.

Acknowledgements

This research is supported by National Natural Science Foundation Project of China (No. 61101084), State Key Laboratory of Networking and Switching Technology Open Project (No. SKLNST-2011-1-02) and the Fundamental Research Funds for the Central Universities.

Author details

Hui Zhang[1,2], Yifeng Xie[1], Liang Feng[1] and Ying Fang[1]

1 Nankai University, China

2 State Key Laboratory of Networking and Switching Technology (Beijing University of Posts and Telecommunications), China

References

[1] Bartomiej Błaszczyszyn, Paul Mühlethaler, Yasser Toor. Stochastic analysis of Aloha in vehicular ad hoc networks, Annals of Telecommunications, Springer-Verlag, 2012.

[2] Franc¸ois Baccelli and Bartłomiej Błaszczyszyn. Stochastic Geometry and Wireless Networks Volume II:Applications, Foundations and trends in networking, Now Publishers, 2009.

[3] Ki Tae Kim, Seong Keun Oh. An Incremental Frequency Reuse Scheme for an OFDMA Cellular System and Its Performance. IEEE Vehicular Technology Conference-Spring, pp.1504-1508, 2008.

[4] M.Haenggi, J.G.Andrews, F.Baccelli, O.Dousse, M.Franceschetti. Stochastic Geometry and Random Graphs for the Analysis and Design of Wireless Networks, IEEE Journal on Selected Areas in Communications, vol.27, no.7, pp.1029-1046, 2009.

[5] Salman Malik, Alonso Silva, Jean-Marc Kelif. Optimal Base Station Placement: A Stochastic Method Using Interference Gradient In Downlink Case, Proceedings of the 5th International ICST Conference on Performance Evaluation Methodologies and Tools, pp.66- 73, 2011

[6] Francois Baccelli and Bartłomiej Błaszczyszyn. Stochastic Geometry and Wireless Networks: Theory, Foundations and Trends in Networking, vol.3, no.3-4, pp.249-449, 2009.

[7] Atsuyuki Okabe, Barry Boots, Kokichi Sugihara, Dr Sung Nok Chiu. Spatial Tessellations: Concepts and Applications of Voronoi Diagrams, Wiley Series in Probability and Statistics, vol.501, John Wiley & Sons, 2009.

[8] François Baccelli, Maurice Klein, Marc Lebourges and Sergei Zuyev, Stochastic geometry and architecture of communication networks, Telecommunication Systems, vol.7, no.1-3, pp.209-227, 1997.

[9] Bartłomiej Błaszczyszyn, Paul Muhlethaler. Stochastic Analysis of Non-slotted Aloha in Wireless Ad-Hoc Networks, IEEE Infocom, 2010.

[10] Hui Zhang, et al. Frequency Reuse Analysis Using Multigraph Theory, International Conference on Green Communications and Networks, 2011.

[11] Jingya Li, Hui Zhang, Xiaodong Xu, et al. A novel frequency reuse scheme for coordinated multi-point transmission. IEEE Vehicular Technology Conference-Spring, pp.1-5, 2010.

[12] Min Liang, Fang Liu, Zhe Chen, et al. A novel frequency reuse scheme for OFDMA based relay enhanced cellular networks. IEEE Vehicular Technology Conference-Spring, pp.1-5, 2009.

[13] Lei Chen, Di Yuan. Generalized frequency reuse schemes for OFDMA networks: optimization and comparison. IEEE Vehicular Technology Conference-Spring, pp.1-5, 2010.

[14] F.Wamser, D.Mittelsta, D.Staehle. Soft frequency reuse in the uplink of an OFDMA network. IEEE Vehicular Technology Conference-Spring, pp.1-5, 2010.

[15] M.Assaad. Optimal fractional frequency reuse (FFR) in multicellular OFDMA system. IEEE Vehicular Technology Conference-Fall, pp.1-5, 2010.

[16] A.Imran, M.A.Imran, R.Tafazolli. A novel self organizing framework for adaptive frequency reuse and deployment in future cellular networks. IEEE International Symposium on Personal Indoor and Mobile Radio Communications, pp. 2354-2359, 2010.

[17] T.Novlan, J.G.Andrews, Sohn Illsoo, et al. Comparison of fractional frequency reuse approaches in the OFDMA cellular downlink. IEEE Global Telecommunications Conference, pp.1-5, 2010.

[18] 3GPP R1-050896, Description and simulations of interference management technique for OFDMA based E-UTRA downlink evaluation, Qualcomm, 2005.

[19] W K Hale. Frequency assignment: theory and applications. Proc. of The IEEE, vol.68, no.12, pp.1497-1514, 1980.

[20] Du Juan, Zhang Yu-qing, Zhan Su-juan. An algorithm to compute the span of the T-Colorings of multigraphs. Journal of the Hebei Academy of Sciences. vol.23, no.3, pp. 1-4, 2006.

[21] 3GPP R1-050738, Interference mitigation considerations and results on frequency reuse. Siemens, 2005.

[22] Cai Wen. Extention Theory. Beijing: Science Press, pp.69-77, 2003.

[23] Victor DeMiguel, Huifu Xu. A Stochastic Multiple-Leader Stackelberg Model. Operations Research, vol.57, no.5, pp.1220-1235.

Chapter 3

Small LTE Base Stations Deployment in Vehicle-to-Road-Infrastructure Communications

Luca Reggiani, Laura Dossi,
Lorenzo Galati Giordano and Roberto Lambiase

Additional information is available at the end of the chapter

http://dx.doi.org/10.5772/55430

1. Introduction

Rational and efficient handling of vehicular traffic and people movement is already nowadays a huge challenge that increasingly needs to be supported by dedicated technologies. Scenarios like Smart Cities and Smart Roads are therefore among the most promising areas in which major technological breakthroughs are expected in the next two decades. The delivery of end user services in these scenarios surely call for an ensemble of different technologies to be displaced together; for successful deployment of such technologies and for positive and effective impact on daily life, one of the most important factors is certainly the efficient, fast and flexible distribution of information. The possibility to transmit a large quantity of data is of course necessary in these kind of scenarios, but is not to be separated from the possibility to have higher spectral efficiency and higher flexibility of spectrum configuration. Another key factor is the cost of the network infrastructure, which can easily become a showstopper even for the most promising service or technology. In fact in recent years, vehicle-to-vehicle (V2V) and vehicle-to-road-infrastructure (V2I) communications have attracted great interest for the potentially extended usage in traffic applications and emergency situations. While in the former case the communication system is realized by means of an ad-hoc network (denoted as VANET), in the latter vehicles, i.e. mobile terminals, exchange data packets each other and with the Internet by means of a base station (BS) situated close to the road side. So, in V2I the BS controls the network and manages the connections with and between the vehicles, simplifying the vehicle transceiver and improving performance, including the reliability of emergency communications.

Obviously the cost of the V2I infrastructure is higher and this aspect should be evaluated carefully w.r.t. the performance advantages and the revenue growth for the operators. One of the objectives of the present work is to show how such a network infrastructure can be already deployed using equipments available for mass production with a high degree of maturity and acceptable costs, thus making Smart Roads deployment economically attractive for operators.

Two solutions are currently object of interest for the V2V and V2I communications: IEEE 802.11p, which is an extension of the IEEE 802.11 standard for local area networks, and the fourth generation LTE (Long Term Evolution) cellular radio system [1–4]. Currently 3GPP LTE system is emerging as the dominant radio access technology for the next 10 to 15 years. It addresses effectively the issues briefly depicted above as it provides effective broadband access and distribution with high spectral efficiency and relatively simple network structure. At the same time, it preserves key features such as high speed mobility, voice services, quality of service, minimal latencies, and high flexibility in the allocation of radio spectrum resources to final users.

Additionally, with reference to the deployment of broadband technologies and therefore also in the case of LTE system, it is nowadays clarified that the role of small cells will be more and more important in order to guarantee real broadband availability for the end users and an higher degree of adaptability of the network to the real geographical and users' density constraints [5, 6]. In other words, next generation networks will have a macro cells layer with the role of making available to the end users the generic access and a micro/small cells layer with the task to take over the higher demand for data rate and coverage in strategic hot spots.

In general, small cell diameter ranges from a few hundred meters to a few kilometers, therefore, for the deployment of a V2I network, it is necessary to locate a quite large number of base stations along the road. Major consequence is that the cost of the base station is to be kept as low as possible. The most important hardware components for this kind of base station are the following:

- One generic processor to handle all Radio Resource Control (RRC), Radio Resource Management (RRM), Operation and Maintainance (OM) and Self Organizing Networks (SON) functionalities.
- Digital Signal Processing (DSP) power to handle Physical layer (PHY) and layer 2 Medium Access Control (MAC), Radio Link Control (RLC) and Packet Data Convergence Protocol (PDCP) functionalities. It can be achieved using multicore devices or an equivalent System-on-Chip (SoC).

This kind of configuration is already adopted in several commercial solutions and several examples of the adopted chips can be found in [7–9]. Typically they support LTE 3GPP releases 8 and 9 [10] up to 2x2 MIMO transmission data rate which is up to 150/75 Mbps, respectively for Downlink and Uplink (DL/UL), for $BW = 20$ MHz channelization. The adoption of DSP processors is usually preferred because it eases software development and upgrades. The indicative cost for such equipment including Radio Frequency (RF) board ranges from 1000 to 2500 Euros considered in volumes, which is in line with the need of mass deployment. The whole small base station can be mounted on any existing support, such as light lamps, traffic signals or similar, so that the cost of deployment of a dedicated mechanical infrastructure is massively reduced. This is feasible because the physical dimension of the base station described can be easily in the range of a shoes box without the need to face with special hardware design issues.

http://dx.doi.org/10.5772/55430

In this work we study the application of LTE small base stations on roads characterized by high traffic density, as for example parts of national highways in the proximity of important cities. Small BSs respond to the necessity of obtaining excellent performance without high capital investments and with easy installment (e.g. on the already existing lattices for traffic signaling) and fast auto-configuration. It should be noted that cellular networks, whose infrastructure is already widely deployed, have been proposed several times for vehicular applications especially in the third generation radio network (UMTS). Nevertheless in LTE and in future next generation radio network, heterogeneous and scaled radio stations, covering overlapping areas of different sizes, will be one of the keys for an optimal, from a technical point of view, and cost effective usage of the bandwidth. So next generation LTE small BSs, specifically designed and dedicated only to the road coverage by means of small cells, could constitute a valid, inexpensive option at least for road scenarios characterized by high traffic and user densities.

The main peculiarities of vehicular communications are the considerable high speed of vehicles and the very low latency that must be guaranteed in high priority emergency data transmissions between the different vehicles through the small cell LTE infrastructure. Examples of emergency services are those to avoid car accidents or dangerous situations before the driver recognition and reaction. Moreover, we may add the widespread necessity of ambitious capacity requirements for future broadband wireless network in order to provide high rate capabilities for audio and video stream applications.

So these networks expose system designers to the challenging compromise between the scarcity of spectral resources, the impairments introduced by radio propagation randomness and the stringent service constraints of vehicle communications in case of emergency. At the heart of this challenge there is the ability to manage radio resources as efficiently as possible in all available dimensions (space, time, frequency or channel, power, modulation and coding).

In this work, we investigate three fundamental features for the physical layer of V2I in LTE small cells: the topological layout, including the role of the antenna directivity function, the scheduling strategies for V2I connection performance and the impact of imperfect channel estimation in fast time varying scenarios.

The chapter is organized as follows: Sect. 2 describes the system model, the possible topological distribution of small base stations in the vehicular communications infrastructures, the simulation assumptions and the analysis strategy to highlight the significant design and performance parameters. Sect. 3 reports the fast adaptive strategies adopted in the analysis. Sect. 4 illustrates the numerical results of the system analysis in terms of the global system parameters previously defined, and finally conclusions are reported in Sect. 5 .

2. System model

We started our investigation on Smart Road scenario considering some specific services which may be of real interest for the end users. Such services are then translated into the respective traffic models and afterwards used as an input for the simulations.

1. *High priority traffic:*
   - Emergency warnings: this service allows a centralized entity to broadcast or multicast to all the affected vehicles the necessary emergency alerts (low data rate consumption) related to the road scenario itself.
   - Emergency calls: this service allows the vehicle to start an emergency call towards public forces, tow trucks, ambulances or also other vehicles moving on the same road.
2. *Medium priority traffic:*
   - Traffic information: by this service the end user automatically receives detailed information about the traffic situation; these data are to be broadcasted to all the vehicles and are handled by a central application server. Then each vehicle can analyze and adopt only the information that are consistent with his planned route, taking decision accordingly. This service is considered a low rate, medium priority broadcast/multicast data transmission.
   - Weather forecast: this service is similar to the previous one but has a different distribution with reference to the geographical regions in which it is broadcasted.
3. *Low priority traffic:*
   - Multimedia services: all the bandwidth remained unused by the previous services will be available for additional services like web browsing, video on demand, VoIP. The transmission of this data is not the primary task of a V2I deployment, but for sure it can bring an added value to the whole infrastructure. For this type of services it is extremely important to apply an efficient scheduling algorithm that may allow a better usage of the remaining available resources.

Having described a bundle of services which are probably the most interesting for the end users, we indicate a number of corresponding LTE features that are to be included in the system to support such services. These important features are then turned into traffic models or general system constraints which are consequently inserted into the simulation work. The goal is to create a link between the real world constraints (given by the network infrastructure, the standard, the services needed by the end users) and our simulation works.

- *Number of supported users*: we hypothesize a small cell would have to support approximately 100 User Equipments (UEs) in the coverage area; these UEs would all be in connected state, i.e. the UE state specified in the standard in which the UE can potentially exchange data with the network and can be paged without requiring the access procedure. Therefore 100 UEs are a reasonable figure for a commercial small cell. This parameter is therefore acceptable and does not represent a limiting factor.
- *Data rate per user*: another key performance figure for a commercial small cell base station is the maximum data rate supported per each user. With a MAC scheduling algorithm of reasonable complexity and channel bandwidth $BW = 20$ MHz, the theoretical limit of the number of users contemporary scheduled on the same subframe is 100 (i.e. the number of the available resource blocks). However a small cell, with reduced complexity and computational power can support around 10 contemporary users. In the low profile scenarios of single antenna or 2 antennas transmission diversity, the whole 75 Mbps data rate allowed by LTE can be divided among the 100 active users, and assuming to serve 10 users per TTI frame, with TTI frame duration of 10 ms, the system allows 750 kbps

constant data rate for each UE. This sounds a very good performance figure allowed with no special demand to the system. The cell size and/or the data rates are obviously to be scaled with the channel bandwidth BW.

- *High mobility receiver*: the base station receiver shall be tailored to sustain at least a user speed of 150 km/h and to avoid interference from the neighboring cells.
- *QoS support*: the support of the different users quality of service is an important key feature and the LTE MAC packet scheduler is the the one in charge of taking decisions on how to deal with the priority and resource allocation.
- *Multicast/broadcast*: these features are already specified in 3GPP release 9 [10] and are thought to dramatically save bandwidth in downlink direction when same information are to be broadcasted or multicasted at the same time to all or a group of users. In this case, in fact, the same data flow is transmitted only once towards many users at the same time and does not need to be repeated: for example, if a given traffic situation feed is to be transmitted at 50 users at the same time, the 50 transmission resources that would be required by unicast mode are reduced to one.

Next subsections show the simulated network deployment, the assumptions on the structure of the radio frame and the assumptions taken for the analysis.

2.1. Small base stations deployment

The application scenario is a network of small cell sites uniformly distributed on the linear road layout, and separated by a distance d_{BS}, each one dedicated to the coverage of a road sector, as sketched in Fig. 1.

The BSs not only provide V2I communications for the vehicles passing in their coverage area, but they also communicate each other either to broadcast the emergency messages or to backhaul Internet information at the BSs directly connected to core network. In fact, in presence of small cells, to maintain low infrastructure and operational costs, we assume that only some BSs are equipped with a dedicated wired or wireless link to the core network and Internet, while n_{BH} BSs between them operate in Relay Node (RN) mode [11]. We assume to perform an 'In-band' relay solution, operating on the assigned system UL/DL frequencies, meaning that links between the base station and the relay nodes are on the same carrier frequency as the link between the base station and the user equipment i.e. the BS-RN link and the BS-UE link are on the same carrier frequency [12]. Each $i-th$ base station is equipped with two antennas:

- its sectorial antenna, that has to cover the road sector and whose radiation pattern $A_{FW}(\theta)$ has to be optimized to provide high coverage levels, despite the multi-cell layout context under investigation possibly impaired by interference due to frequency reuse factor equal to one. The corresponding radio interface communicates with both the associated mobile UEs and the illuminated adjacent $(i+1)$-th BS, transmitting and receiving respectively on the system DL and UL frequencies. The fundamental parameters that define $A_{FW}(\theta)$ are the 3dB beamwidth θ_{3dB} and the beam orientation, set in order to guarantee an external angle θ_{EXT} as defined in Fig. 1.

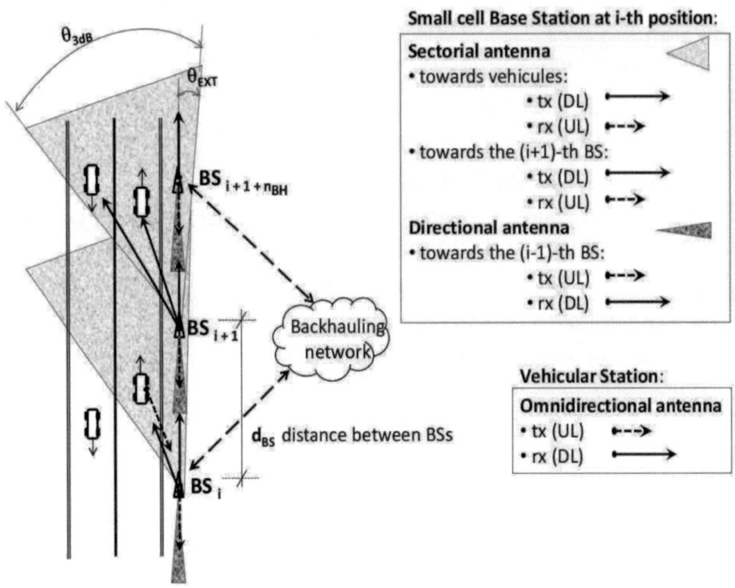

Figure 1. Network topology for smart road applications and vehicular communications. In backward direction, each BS could behave as a UE: for emergency info it transmits on the uplink (UL) and receives on the downlink (DL). For backhauling data it has two alternatives: transmitting on UL or receiving on DL.

- another directional antenna, dedicated to BS to BS (Infrastructure to Infrastructure, or Infra to Infra) communications, so pointing, with a highly selective radiation pattern $A_{BW}(\theta)$ towards the adjacent $(i-1)$-th BS, in backward direction w.r.t. the one illuminated by the sectorial antenna. The radio interface associated to this antenna operates as an UE equipment, transmitting on the UL and receiving on the DL.

This communication system topology is chosen in order to make the emergency information flow propagate either forward on the DL, radiated by the sectorial antenna, or backward on the UL, supported by the directional antenna equipment. This responds to the necessity of broadcasting emergency information with very low latency in both road directions. At the same time, the backhaul flow transport can exploit two alternatives:

- in forward direction, so using the sectorial antenna to transmit in DL, and the directional antenna to receive;
- in backward direction, so using the directional antenna to transmit in UL, and the sectorial one to receive .

The choice between the two solutions must take into account the peculiarities of the two traffic flows and the frame structure type, if *Type*1 FDD (Frequency Division Duplex) or *Type*2 TDD (Time Division Duplex). This topic will be discussed in the next Sect. 2.2.

The V2I communication performance, in the area covered by the sectorial antennas, is simulated selecting, among spatially uniformly distributed vehicles, n_v active mobile users at speed of $v = 130$ km/h in the two road directions at a medium-low inter distance, i.e.

between 25 and 100 m. The gain pattern of the two antennas at the base stations, $A_{FW}(\theta)$ and $A_{BW}(\theta)$, are specified as [13]

$$A(\theta) = -min(12(\frac{\theta}{\theta_{3dB}})^2, A_m) + G; [dB] \quad (1)$$

where θ is in [-180, 180] degrees range, θ_{3dB} is the 3dB beamwidth in degrees, A_m is the maximum attenuation and G is the antenna gain.

Sectorial antenna parameters will be appropriately adjusted in order to cover the overall road area, minimizing the overlapping regions, possibly impaired by co-channel interference for low values of reuse factor r. On the other hand, mobile terminals are equipped with omnidirectional antennas.

Concerning V2V propagation channel models, due to the high relative speeds involved, especially when we consider vehicles coming from opposite directions on the same route that experience relative speed till to 300 km/h, effects such as strong time variance and non stationarity are particularly pronounced and cannot be neglected. However, unlike V2V propagation, the V2I propagation channel model generally shows great similarity to conventional cellular propagation channels, e.g. macro-cells and micro-cells, for which standardized channel models can be adopted [13]. All of these models provide key parameters to generate the appropriate distributions of path-loss and fading statistics, temporal variance and delay spread as particular scenarios require.

It must be mentioned that a typical V2I scenario along a highway with a fixed roadside transmitter and a vehicular receiver has been faced in [14] through a computationally advantageous implementation based on geometrical considerations, but we expect that in this context a more general channel model is more suitable to investigate the performance of the proposed system deployment. In our simulations the input parameters of the 3GPP SCM ('Spatial Channel Model') [15] have been conveniently initialized to simulate the application-specific scenario covered by LTE micro-cells. An SCM *MicroUrban* scenario (as suggested in [16]) has been used to generate path loss, shadowing and fast fading realizations, with either *LoS* or *NLoS* propagation in order to test the system performance under different propagation severity levels.

Antenna and channel model parameters used in the simulations are summarized in Table 1.

2.2. The frame structure

According to the LTE physical layer, we consider a communication system with an Orthogonal Frequency Division Multiple Access (OFDMA)-based frame structure as shown in Fig. 2, where subcarriers and time-slots are grouped in the so-called Resource Blocks (RB), which are the basis for the radio resource assignment to the users.

Each frame has a fixed time duration $T_f = 10$ ms and is divided in 10 subframes of duration $T_{sf} = 1$ ms, each one of them corresponding to 2 OFDMA slots of 0.5 ms. One OFDMA slot contains 7 symbols (with normal cyclic prefix) or 6 symbols (with extended cyclic prefix). On the frequency domain, the bandwidth BW is divided in N_{sch} subchannels. The channelization considered for the small cell deployment under test will be $BW = 1.4$ MHz, $BW = 5$ MHz

Base station equipment	
Sectorial Antenna Gain G[dBi]	$[17 - 30]$
" 3-dB beamwidth θ_{3dB} [°]	$[15 - 60]$
" external angle θ_{EXT} [°]	$[3.75 - 15]$
" Maximum power [dBm]	46
User station equipment	
Antenna Gain [dBi]	0
Radiation Pattern	omnidirectional
Noise figure [dB]	7
Channel model	
Path loss [dB], d [m]	SCM Urban Micro LoS: $30.18 + 26log_{10}(d)$ NLoS: $34.53 + 38log_{10}(d)$
Shadowing [dB]	SCM Urban Micro (LoS: $\sigma_S = 4$; NLoS: $\sigma_S = 10$)
Fast fading	SCM Urban Micro (LoS/NLoS), $v = 130$ km/h
Co-channel Interference model	Explicit (from the co-channel cells)
Cell Layout	
Inter-cell distance d_{BS} [m]	$[250 - 1500]$

Table 1. Antenna and channel model parameters.

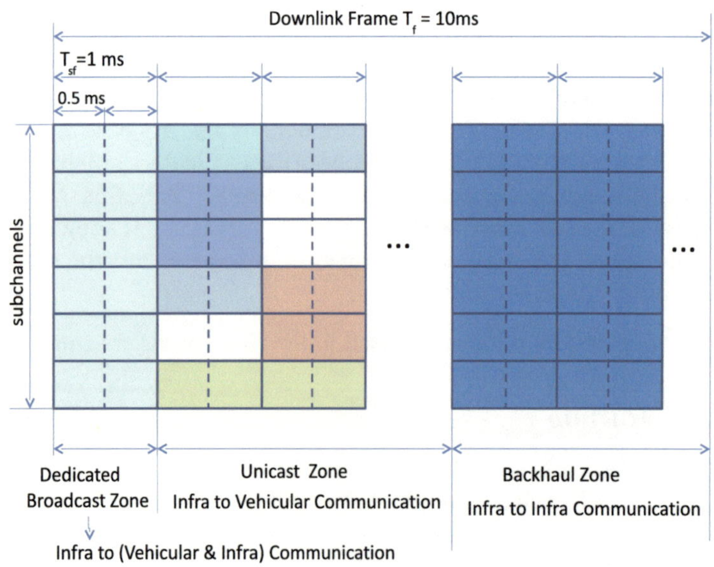

Figure 2. Structure of the downlink frame with three zones dedicated to the following types of traffic flows: Emergency Broadcast information, Unicast communication and Backhauling data produced by the BSs that are not directly connected to the core network.

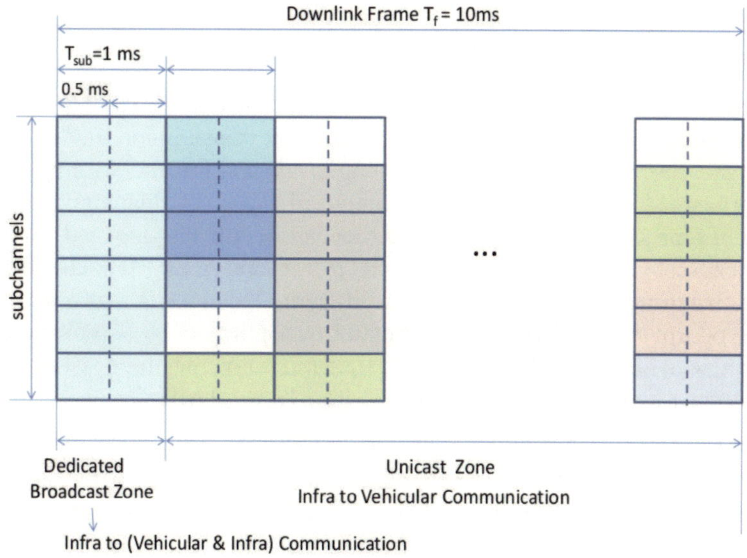

Figure 3. Structure of the downlink frame with two zones, the Emergency Broadcast Zone and the Unicast Zone, assuming that the backhaul flow is transmitted on the UL. Part of the Unicast zone is left free (blank spaces) for respecting the limits on the backhauling capacity when $n_{BH} > 0$.

and $BW = 10$ MHz corresponding to 6, 25 and 50 subchannels. Each subchannel is always constituted by 12 subcarriers. The minimum time-frequency portion to be allocated to a user is one RB, i.e. 1 subframe x 1 subchannel, providing a modulation rate of 168 or 144 symbols/ms, respectively for normal and extended cyclic prefix.

In the LTE standard two types of frame structure are defined, Type 1 FDD and Type 2 TDD. Both of them could be considered, but we concentrated on the FDD frame structure. The performance results discussed in the sequel will refer to the downlink since it is the most stressed and critical direction in this application. In this context, according to the types of traffic defined in Sect. 1, we assume the following allocation rules:

- emergency information, with the highest priority are allocated in the broadcast zone;
- medium and low priority traffic towards vehicles is allocated in the unicast zone according to a smart algorithm for maximizing the overall capacity on this portion of the time-frequency OFDMA frame. As explained in Sect. 4, this portion will be subject also to limits for respecting the backhauling capacity.
- backhaul information is allocated, with high priority, in portions of the time-frequency OFDMA frame (UL or DL) that can be changed adaptively in each new $T_f = 10$ ms.

We observe also that, while emergency and unicast information are communications between BSs and vehicles, backhauling information are intra-communications, from BS to BS.

As previously introduced in Sect. 2.1, and shown in Fig. 1, the backhauling flow could be transmitted either on the UL frame (using the directional antenna) or on DL frame (using the sectorial antenna). The adoption of one of the two solutions strongly depends on the capacity and congestion of the two UL and DL directions. These two alternatives provide two different fragmentation of the DL frame, as reported in Figs. 2,3.

In both cases, the key performance factor is the satisfaction of the DL emergency data transmission from BS to the vehicles. The most efficient transmission mode for emergencies information are the broadcast resources being available for the Multimedia Broadcast Multicast Service (MBMS) from 3GPP LTE standard [10]. In this configuration, a single dedicated subframe is reserved on all over the bandwidth (as depicted in Fig. 2). The dedicated broadcast channel covers all the RBs (6 for $BW = 1.4$ MHz channelization) in a subframe, transmitting with a fixed rate of 1 bit/sym. This means that a single emergency message (i.e. occupying one RB per frame) could transport 168 bit/s (normal cyclic prefix), and the ensemble of highest priority messages transmitted on the broadcast zone corresponds to a maximum of 1 kbit/s (always on a bandwidth $BW = 1.4$ MHz).

The second zone in Figs. 2,3 concerns the downlink transmission of unicast traffic flows from BS to vehicles. This is a contention zone, in which Resource Blocks assigned to a user are processed according to its channel quality, benefiting of the adaptive modulation capability included in the OFDMA standard. Adaptive modulation, in fact, allows to choose for each subchannel the most efficient modulation and coding scheme (MCS) supported by the Signal-plus-Interference-Noise-Ratio (SINR) experienced on that subchannel. The modulation and coding profiles considered in the simulations are 5, as reported in Table 2, starting from the most robust one that provides $\eta = 1$ bit/sym rate to the most efficient one corresponding to $\eta = 6$ bit/sym rate. Even if LTE standardizes till to 14 different MCS, it has been verified [17] that such a fine granularity is generally not necessary and a lower number of profiles performs equally well on mobile channels with multiple cells. We emphasize again that, unlike the transmissions of unicast information, the broadcast transmissions of the safety messages are protected using the most robust MCS. Note that, in case of a unicast traffic load less than 100 percent, some RBs can remain not allocated (the blank blocks in the figures), either for not overloading the intra-BSs backhaul or for decreasing the interference realizing a sort of fractional reuse. The role of the resource allocation algorithms (Sect. 3) will be fundamental for managing the time and frequency positions of the not allocated RBs in interfering cells in order to reduce co-channel interference in a multi-cell scenario. The third transmission zone, when present (Fig. 2), is dedicated to the transport of the backhauling flow, activated only for the BSs not directly connected to the Internet: if $n_{BH} = 0$ the backhauling zone is not requested, while for $n_{BH} \geq 1$ the backhauling zone has to be dimensioned at every BS not connected to Internet, considering that the backhauling capacity requirement will increase from BS to BS between two consecutive Internet connections. If the backhaul flow is transmitted on the UL frame, the downlink frame will be segmented into two zones (Fig. 3): more space can be dedicated for the unicast zone in DL, while UL frame will host the backhauling traffic. Considerations on the frame structure type, on the expected volume of backhauling traffic, on the deployment parameter n_{BH}, that influences the backhauling traffic volume, must be taken into account in order to make a correct allocation of the RBs during the BSs operations. The numerical simulations presented in Sect. 4 aim at highlighting the performance limits for a realistic range of the system parameters, summarized in Table 2.

Downlink Air Interface Parameters	
Center Frequency [GHz]	2.0
Bandwidth [MHz]	[1.4, 5, 10]
Number of subcarriers per RB	12
Number of RB / time-slot	[6, 15, 25]
MCS rates [bit/sym]	1, (4 x 1/2), (4 x 3/4), (6 x 3/4), (6 x 4/4)
Frame duration [ms] T_f	10
Subframe duration T_{sf} [ms]	1

Table 2. LTE compliant system parameters.

2.3. Design parameters and performance evaluation analysis

An analysis of the capability of the proposed small base stations layout for a typical smart road, constituted e.g. by a highway, will be developed according to the following phases:

- Definition of a typical traffic scenario between a vehicle and a base station situated at the road side.
- Definition of the parameters for the base stations setup and deployment.
- Definition of the output performance measures of the network.

The traffic scenario will be constituted by a typical highway, of width equal to 24 m and vehicles at constant velocity of 130 km/h. The communication traffic is measured by the number of n_v connected vehicles per 100 m in the highway (Fig. 1). We assume that an emergency broadcast subframe is reserved for each DL time-frame (corresponding to 1/10 of the OFDMA resources) and each active vehicle requires $k_u = 1$ resource blocks per time-frame in the unicast zone for its medium priority traffic. Note that an increase in the number of active vehicles n_v, i.e. the number of RBs to be allocated on the unicast zone, causes also an increase in the backhauling traffic. In particular, one of the limits imposed on the network is given by the amount of traffic that the BSs can transmit on the intra-BSs backhauling without overloading and data losses. We remark that, in the proposed deployment, part of the backhauling will be performed by the same BSs acting as relays (Fig. 1): we have assumed that, when $n_{BH} > 0$, each BS can occupy only a fraction $1/(n_{BH}+1)$ of the time-frame portion that remains after the broadcast subframe allocation (i.e. 9/10 of the time-frame). This rule prevents from the overloading of intra-BSs backhaul, that is implemented in the UL space (for example in the UL time-frame for the FDD case). As already observed, this unoccupied portion of the DL bandwidth is changed dynamically according to the number of connected vehicles and to their channel conditions.

The deployment analysis will be performed focusing on the following design parameters: the distance d_{BS} between base stations, the number n_{BH} of base stations between two connections with the Internet core network, the sectorial antenna beamwidth θ_{3dB} and external angle θ_{EXT}.

Finally the output performance will be measured by the following parameters:

- Vehicular density n_v, as the number of active n_v vehicles in 100 m simultaneously served (medium priority traffic) with a given outage probability.

- Overall capacity [Mbps] for low priority traffic.

3. Scheduling and allocation techniques for vehicular communications

In OFDMA systems like LTE, smart scheduling and allocation of radio resources are crucial aspects for optimal performance. The allocation techniques are based on the knowledge of the channel state indicators (CSI) for each user and for each subchannel. According to the procedures in LTE standard the updating rates of CSI reports from the users should be comparable with the coherence time of the channel and compatible with the maximum latency admitted in emergency data exchange. This is even more important in this kind of application, characterized by requirements of high priority and low latency of emergency communications. This challenging factors combination constitutes the main limiting factor in these applications [3, 4]. Smart resource scheduling and allocation have algorithm updating rates that are affected not only by the periodicity of the transmission of CSI reports (the minimum admissible period time in LTE is 1 ms), but also by the computing time of the algorithm itself. As we are considering high speed terminals, unforeseen channel variations during the algorithm updating time could cause imperfect algorithm allocations due to imperfect estimation of channel gains. The imperfect estimations of channel state indicator have to be taken into account, since they are expected to have an impact on the effectiveness of scheduling strategies, and great attention should be paid on the refinement of fast updating procedures. The power and channel adaptation approach proposed in [18] showed, in an interference limited multi-cell scenario with high speed mobile users, high performance, simplicity and the fast adaptivity required in the V2I deployment.

We emphasize again that the smart algorithm is not applied for the allocation of the short emergency messages, as this last service can benefit of the dedicated broadcast channel on the whole bandwidth, but for the assignment of RBs to the users on the unicast zone in order to increase the spectral efficiency of the communications between BS and the vehicles so exploiting the benefits of multi-user diversity.

Finally, the smart allocation of RBs, which comprises also the power adaptation, will be crucial for limiting the interference in a multicellular scenario with a low frequency reuse factor, as desirable in cost-efficient network solutions. In this context, the portions of the time-frame not assigned to any vehicle, to avoid overloading the intra-BSs backhauling (blank spaces in Fig. 1), will decrease the co-channel interference also in overlapping coverage areas served by BSs operating all at the same frequency.

4. Numerical results

In this Section, numerical results will be focused on the final coverage, i.e. the number of connected vehicles, the outage probability and the throughput for different selections of the cell parameters (d_{BS} and n_{BH}), algorithm parameters and channel models, including the effect of imperfect channel estimation.

The simulated network is composed by a line of 21 consecutive equi-spaced BSs (d_{BS}). The road used in the simulations has a width equal to 24 m.

As introduced in Sect. 2.2, the backhauling flow among base stations has been realized in the FDD UL bandwidth. Consequently the capacity results are collected in the DL, according to the frame structure in Fig. 3. As already observed in Sect. 2.3, this choice requires a limit on the downlink frame occupation for not overloading the backhauling allocations.

In the next Sect. 4.1, we present the results on the sector configuration in terms of antenna parameters and distance between adjacent base stations in order to select the most appropriate geometric parameters for the V2I scenario. Here the numerical results are derived in terms of average SINR (Signal-to-Interference Noise Ratio), i.e. without the contribution of shadowing and fast fading. Then the two dynamic Micro Urban scenarios have been tested, LoS and NLoS respectively. So Sect. 4.2 and 4.3 report the capacity results, expressed by the average density of vehicles that can be served with medium priority and the available low priority throughput, obtained averaging the capacities of the 21 sectors. The allocation of the radio resources is realized by the following steps: first the RBs dedicated

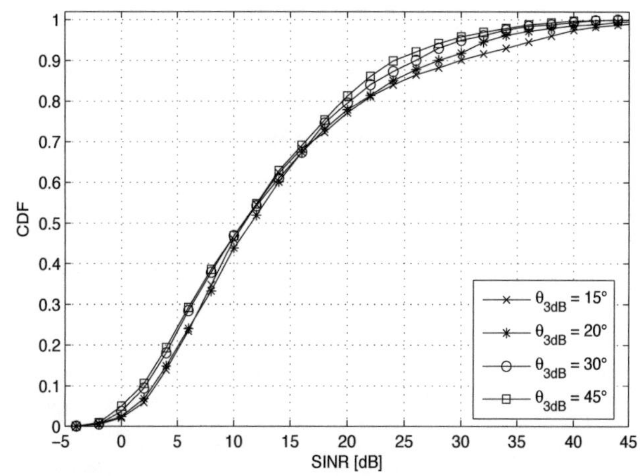

Figure 4. Cumulative distribution function of the average SINR for uniformly distributed vehicles in the 21 cells layout (LoS path loss model, $\theta_{EXT} = 7.5°$and $d_{BS} = 500$ m).

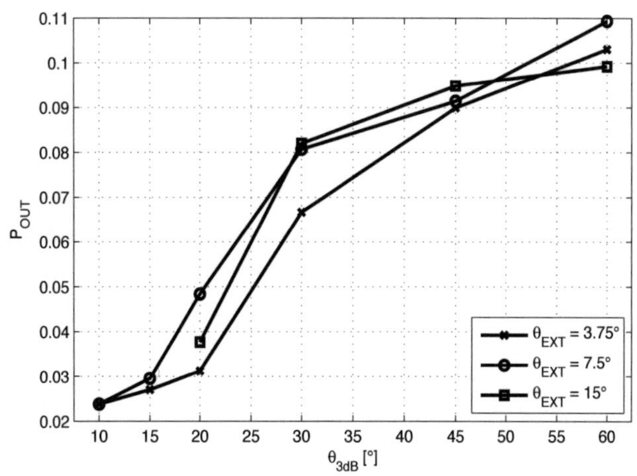

Figure 5. Outage probability as a function of θ_{3dB} for uniformly distributed vehicles in the 21 cells layout and different θ_{EXT} (LoS path loss model and $d_{BS} = 500$ m).

to the broadcast transmission of high priority emergency communications are allocated in each time-frame, then a single RB is assigned to each medium priority data request. This last value returns the number of vehicles that can exchange medium priority data, such as road and traffic conditions. These medium priority RBs can transmit using one of the available modulation profiles, according to the SINR at the vehicle. The resource allocation is performed by means of algorithms that exploits the multi-user diversity, represented by the diversity indicator I_D [18]. Finally the rest of the available bandwidth is used for the low priority Web and Multimedia services. A part of the RBs in the DL frame is left free because of (i) the necessity of not overloading the bandwidth required for the backhauling and (ii) the necessity of realizing a sort of fractional reuse in order to limit the interference among base stations. In fact, we assume that all the base stations use the same frequency achieving a potential reuse factor equal to 1. In the results presented here, the bandwidth is used for a portion between 0.4 and 0.7 according to the parameter n_{BH}, i.e. the number of base stations that are required to accumulate and forward the backhauling traffic.

Except for the results in Fig. 10, a fading margin FM equal to 10 dB has been always included in the SINR computation during the RBs assignment process in order to limit the impact of channel estimation errors, interference and fading variations.

4.1. Sector configuration

The study of the sector configuration is focused on two fundamental parameters of our scenario and it is preliminary for the successive capacity results. Since this particular scenario is characterized by a line of consecutive base stations, we are interested in the antenna beamwidth and the base stations inter-distance that are compatible with the BS transmit power and the received SINR at the locations of vehicles moving along the road. For this reason we have computed the average measured SINR along the road and the outage probability according to the space loss propagation for different distances between the vehicles and the BSs locations.

The antenna aperture is assumed as the 3 dB beamwidth (θ_{3dB}) in the 3GPP directivity function and the antenna gain is set according to this variable beamwidth. Here the particular layout of the small cell, that should cover a portion of the road, requires a narrow antenna beam for limiting the interference impact on other cells operating at the same frequency. In addition, each antenna covers also the next BS for backhauling purposes and the boresight is set in order to guarantee an external angle equal to θ_{EXT}, as shown in Fig. 1. The geometrical layout, the antenna and the path loss models are used for computing the average SINR experienced by all the vehicles uniformly distributed on the road; each vehicle is assigned to the cell, or base station, that guarantees the best average $E[SINR]$ and the outage P_{OUT} is defined as the probability that $E[SINR]$ is below the lowest threshold profile (Table 2).

In Fig. 4 it is possible to see the $E[SINR]$ cumulative distribution function obtained with different θ_{3dB}; results are similar for the NLoS path loss model and for different BS distances $250 < d_{BS} < 2500$ m. We may observe that (i) $E[SINR]$ does not depend significantly on d_{BS} at the distances under analysis, since the system is interference limited (not noise limited), and (ii) $E[SINR]$ improves slightly for θ_{3dB} approaching 15°. This is confirmed by Fig. 5 where P_{OUT} is reported as a function of different θ_{3dB} and θ_{EXT}. According to these geometrical results, a beamwidth $\theta_{3dB} = 15°$with a boresight that guarantees $\theta_{EXT} = 7.5°$is chosen for the capacity simulations of the next Sections.

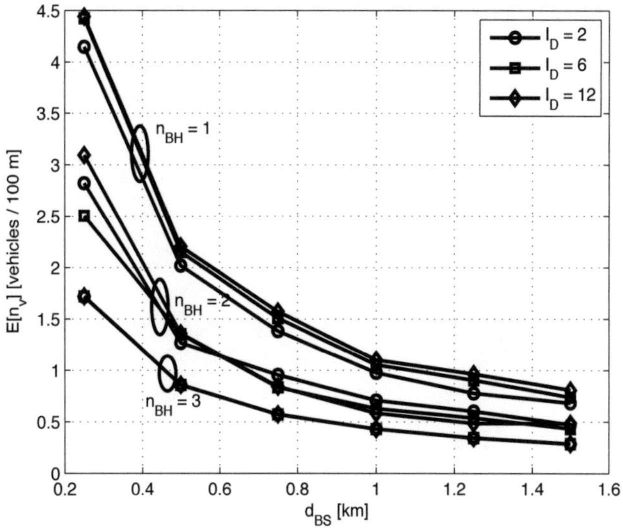

Figure 6. Average vehicles density supported in each subframe T_{sf} as a function of BSs distance for different n_{BH} in the 21 cells layout (SCM LoS channel model, 5 MHz).

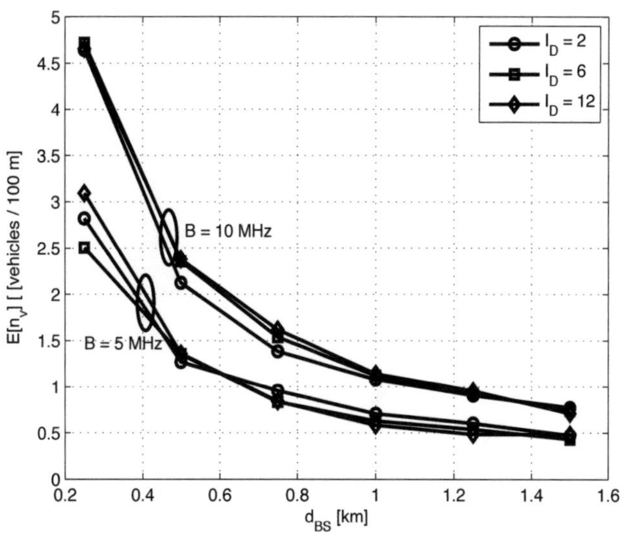

Figure 7. Average vehicles density supported in each subframe T_{sf} as a function of BSs distance for different bandwidths in the 21 cells layout (SCM LoS channel model, $n_{BH} = 2$).

4.2. Scenario 1: SCM Micro Urban LoS

Numerical results are obtained averaging 100 independent fast fading realizations of the SCM Micro Urban LoS and NLoS channel models and each realization represents the evolution of the channel on 10 time-frames for a total time equal to 100 ms. For each realization, it is simulated the behavior of the system in each of the 100 consecutive subframes and in each of the 21 sectors taking into account, obviously, the mutual interference generated on

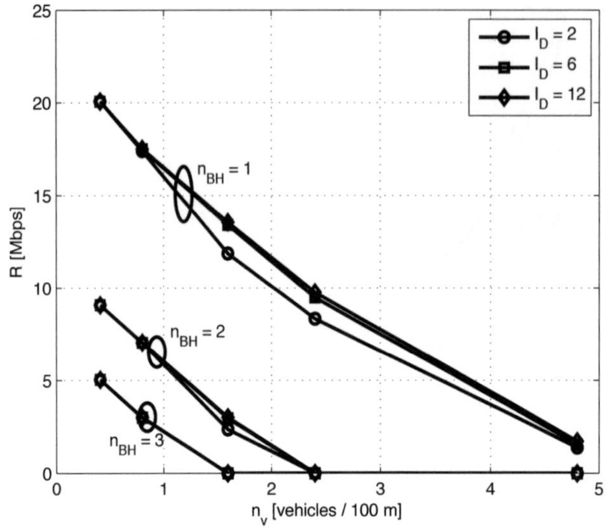

Figure 8. Average low priority throughput as a function of the vehicles density for different n_{BH} in the 21 cells layout (SCM LoS channel model, $d_{BS} = 500$ m, 5 MHz).

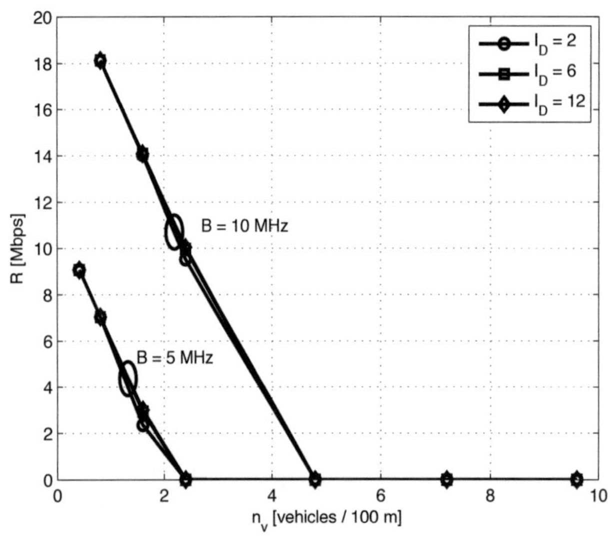

Figure 9. Average low priority throughput as a function of the vehicles density for different bandwidths in the 21 cells layout (SCM LoS channel model, $d_{BS} = 500$ m, $n_{BH} = 2$).

the receivers side by all the 21 downlink signals. We remark that the effective reuse factor is greater than 1 since only a fraction of the available RBs is reserved to data traffic for not overloading the backhauling. We observe that, if the fraction of used RBs is $1/(n_{BH}+1)$ (Sect. 2.3), the effective reuse factor is equal to $n_{BH} + 1$; however we remark that this bandwidth portion is not static but it is selected dynamically according to the SINR experienced by the vehicles and the number of connected vehicles. High priority traffic for emergency situations

http://dx.doi.org/10.5772/55430

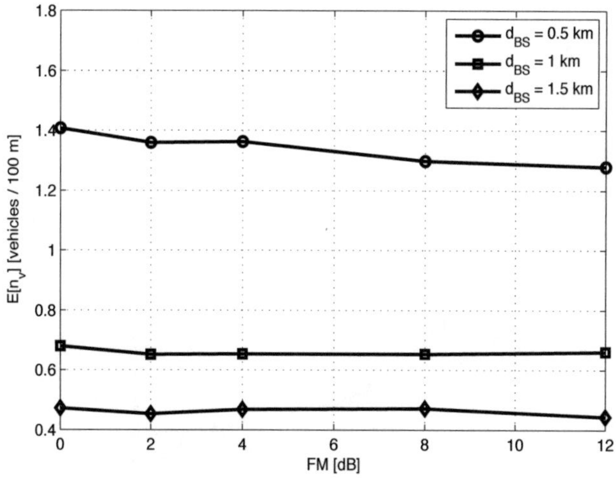

Figure 10. Average vehicles density in each subframe T_{sf} as a function of the fading margin applied to the modulation profile thresholds for increasing robustness against channel estimation errors (SCM LoS channel model, 5 MHz, $n_{BH} = 2$).

is assigned to the broadcast channel, composed by the subframe (all the RBs) at the beginning of each time-frame.

Medium priority traffic is solved by assigning a single RB to the connected vehicles. This assignment is done according to the SINR conditions at each user and hence exploiting the multi-user diversity. The simulation results refer to an outage probability P_{NA} defined here as the probability of not being assigned any RB in a generic subframe. This measure constitutes an upper bound for the probability that a vehicle will not be assigned a RB within the coherence times of channel and interference.

Fig. 6 reports the average number $E[n_v]$ of vehicles per 100 m and per each subframe as a function of d_{BS} and respecting the constraint of receiving an RB for medium priority traffic and a $P_{NA} = 0.01$: it can be observed how the decrease of the density $E[n_v]$ is mostly given by the increasing value of d_{BS}. In the same Fig. 6 it can be appreciated the impact of the wired backhauling access period n_{BH} on $E[n_v]$ while, in Fig. 7, the impact of the bandwidth. If we assume that the time period between two allocations for the same vehicle is equal to N subframes (for example $N = 10$ for a time-frame), the values of $E[n_v]$ reported in the graphs should be multiplied by N for obtaining the number of vehicles, in 100 m, that can be connected and served by the system. So, from these results, it is also possible to appreciate the maximum delay acceptable between two consecutive medium priority RBs.

Low priority traffic is assigned, in a best effort way, to the remaining available RBs after the allocation of the broadcast channel and all the RBs for medium priority traffic. The amount of the traffic available for Web browsing and Multimedia services is represented by the available residual capacity. Figs. 8 and 9 report the low priority traffic as a function of the density n_v of connected vehicles for different n_{BH} and bandwidth values. It is clear that low priority traffic is available when the number of RBs assigned with medium priority is lower than the maximum compatible with the available bandwidth.

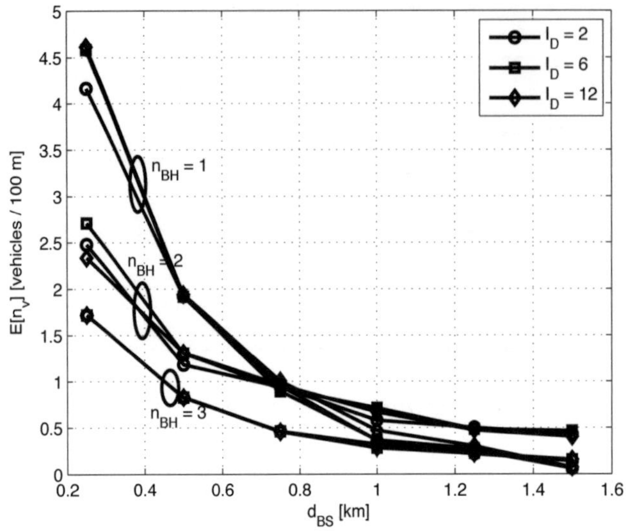

Figure 11. Average vehicles density in each subframe T_{sf} as a function of BSs distance for different n_{BH} in the 21 cells layout (SCM NLoS channel model, 5 MHz).

We can also observe that multi user diversity, represented by the factor I_D, has not a great impact on performance since its effect is limited by the fairness mechanisms introduced for guaranteeing the maximum number of vehicles connected with just one medium priority RB.

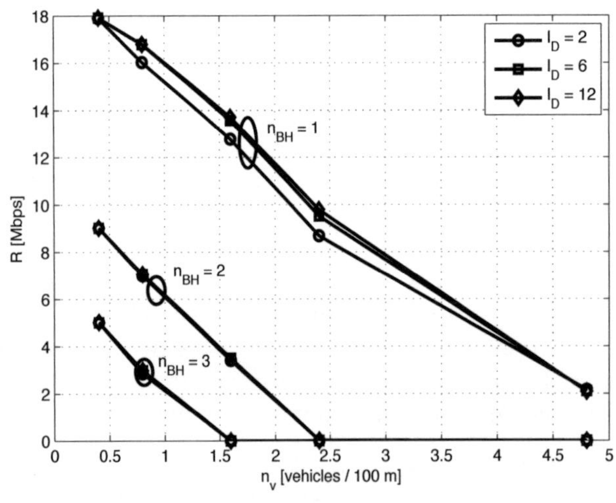

Figure 12. Average low priority throughput as a function of the vehicles density for different n_{BH} in the 21 cells layout (SCM NLoS channel model, 5 MHz).

Finally, Fig. 10 is dedicated to the evaluation of the fading margin impact on the performance: fading margin is used against the variations of interference and channel, including estimation errors, w.r.t. the CSI used at the BSs for the allocation procedures. We observe that a fading margin around 10 dB, generally sufficient for including the channel estimation imperfections, does not affect seriously the density of vehicles that are served in the system. This result is due to the values of SINR at the vehicles, which are generally high w.r.t. the low modulation profiles that can be assigned to the RBs with medium priority.

4.3. Scenario 2: SCM Micro Urban NLoS

Results obtained for the SCM Micro Urban NLoS channel model are equivalent to those presented in the previous Sect 4.2. From Figs. 11 and 12 it may be observed that, as expected, (i) performance is slightly worse than in the LoS case and (ii) performance decreases more severely as d_{BS} increases.

5. Conclusions

In this work we have studied the deployment of LTE small base stations along roads characterized with high traffic density in order to provide vehicle-to-infrastructure (V2I) communication services. Results have been provided to address the V2I deployment and the set-up of the geometric configuration distinguishing different classes of traffic in the system, from high priority, associated to emergency situations, to medium and low priority services like weather forecast, traffic situation and other multimedia services. Numerical results show that the system, with a proper use of the LTE relay capabilities for the backhauling between adjacent base stations, can provide interesting results for a cost effective deployment of vehicle-to-infrastructure communications.

Author details

Luca Reggiani[1], Laura Dossi[2],
Lorenzo Galati Giordano[3] and Roberto Lambiase[3]

1 Dipartimento di Elettronica ed Informazione, Politecnico di Milano, Milano, Italy
2 IEIIT-CNR, c/o Dipartimento di Elettronica ed Informazione, Politecnico di Milano, Milano, Italy
3 Azcom Technology s.r.l., Rozzano, Milano, Italy

References

[1] G. Karagiannis, O. Altintas, E. Ekici, G. J. Heijenk, B. Jarupan, K. Lin, and T. Weil, "Vehicular networking: a survey and tutorial on requirements, architectures, challenges, standards and solutions", IEEE Commun. Surveys and Tutorials, vol. 13, no. 4, 2011.

[2] M. Kihl, K. Bur, P. Mahanta, E. Coelingh, "3GPP LTE downlink scheduling strategies in vehicle-to-Infrastructure communications for traffic safety applications", 17th IEEE Symposium on Computers and Communication, 2012.

[3] A. Vinel, "3GPP LTE Versus IEEE 802.11p/WAVE: Which Technology is Able to Support Cooperative Vehicular Safety Applications?", Wireless Communications Letters, IEEE , vol.1, no.2, pp.125-128, April 2012.

[4] S-Y. Pyun, D-H. Cho, J-W. Son,"Downlink Resource Allocation Scheme for Smart Antenna based V2V2I Communication System", IEEE Vehicular Technology Conference, Sept. 5-8, 2011.

[5] J. Hoadley, P. Maveddat, "Enabling small cell deployment with HetNet", Wireless Communications, IEEE, vol. 19, no.2, pp.4-5, April 2012.

[6] B. A. Bjerke, "LTE-advanced and the evolution of LTE deployments", Wireless Communications, IEEE , vol.18, no.5, pp.4-5, October 2011.

[7] Cavium multi-core architecture with LTE/3G hardware accelerators and digital front end (DFE) specifications, available at http://www.cavium.com/OCTEON-Fusion.html

[8] Texas Instruments Multicore Fixed and Floating Point System on Chip specifications, available at http://www.ti.com/product/tms320c6670

[9] Texas Instruments ARM Cortex-A8 processors specifications, available at http://www.ti.com/lsds/ti/arm/sitara_arm_cortex_a_processor/... sitara_arm_cortex_a8/overview.page

[10] 3GPP specifications, available at http://www.3gpp.org/specifications,last accessed Jan.12, 2012.

[11] 3GPP TR36.806 V9.0.0: "Evolved Universal Terrestrial Radio Access (E-UTRA); Relay architectures for E-UTRA (LTE-Advanced)", 2010.

[12] C. Coletti, P. Mogensen, R. Irmer, "Deployment of LTE In-Band Relay and Micro Base Stations in a Realistic Metropolitan Scenario," Vehicular Technology Conference (VTC Fall), 2011 IEEE , vol., no., pp.1-5, 5-8 Sept. 2011.

[13] C. F. Mecklenbrauker, A. F. Molisch, J. Karedal, F. Tufvesson et al., "Vehicular Channel Characterization and Its Implications for Wireless System Design and Performance", Proceedings of the IEEE, no. 7, Jul 2011, pp. 1189 - 1212.

[14] N. Czink, F. Kaltenberger, Y. Zhou, et al., "Low Complexity Geometry-based Modeling of Diffuse Scattering", Proc.Eur. Conf. Antennas Propagation, Barcelona, Spain, Apr. 12-16, 2012.

[15] 3GPP TR 25.996 V9.0.0, "Spatial Channel Model for Multiple Input Multiple Output (MIMO) simulations (Release 9)", 2009-12.

[16] H. Boeglen, H. Benoit, L.Pascal, et al., "A survey of V2V channel modeling for VANET simulations", IEEE Eighth International Conference on Wireless on-Demand Network Systems and Services, Jan. 26-28, 2011.

[17] L. Galati Giordano, L. Reggiani, L. Dossi, "Discrete-rate allocation schemes in multi-cell scenarios with power adaptation",IEEE 21st International Symposium on Personal Indoor and Mobile Radio Communications (PIMRC), Sept. 26-30, 2010.

[18] L. Reggiani, L. Galati Giordano, L. Dossi, "Fast Power and Channel Adaptation for Mobile Users in OFDMA Multi-Cell Scenarios", Chapter 8 in "Vehicular Technologies: Increasing Connectivity", INTECH, Apr. 11, 2011.

Chapter 4

Deploying ITS Scenarios Providing Security and Mobility Services Based on IEEE 802.11p Technology

Pedro Javier Fernández Ruiz,
Fernando Bernal Hidalgo, José Santa Lozano and
Antonio F. Skarmeta

Additional information is available at the end of the chapter

http://dx.doi.org/10.5772/55285

1. Introduction

It was several years ago when the importance of vehicular communications rapidly grew. The research community on Intelligent Transportation Systems (ITS) had been working for years on autonomous systems focused on either the infrastructure or the vehicle side. This fact is still evident in current systems for traffic monitoring, safety or entertainment integrated in commercial vehicles. Nonetheless, this market inertia is planned to gradually change in the short term, due to the vast amount of research in vehicular communications and cooperative systems that has appeared in the last years. According to new schemes, infrastructure and vehicle subsystems will not be independent anymore. Communication networks should interconnect infrastructure processes (I2I - infrastructure to infrastructure); they should make easier the provision of services to vehicles (V2I/I2V - infrastructure to vehicle); and they should be the seed of future cooperative services among vehicles (V2V - vehicle to vehicle).

As a result of the great research efforts on vehicular communications we are now immersed in the phase of developing previous theoretical or simulated advances and getting preliminary results [1]. The European Union is aware of this necessity and the Sixth and, above all, the Seventh Framework Program calls have been especially focused on field operational tests (FOT) projects, such as the German simTD, the French SCORE@F, the Spanish OASIS, or the recent European DRIVE C2X and FOTsis. Although these initiatives start from the basis of previous research projects, such as CVIS or Coopers, preliminary developments made on those projects should be further extended to obtain more complete communication stacks necessary to perform a wide set of tests in FOTs [2]. Due to that, the European Union agreed to found a project like IPv6 ITS Station Stack for Cooperative ITS FOTs (ITSSv6), whose main aim is to comform an IPv6-based communication stack ready to be used by FOT projects. Part of the work presented in this chapter has been carried out inside ITSSv6 and it has been in this frame where we have realized the great lack of security countermeasures currently available for FOT-equivalent evaluations.

Standardization efforts in ISO TC 204, based on the Communications Access for Land Mobiles (CALM) concept, and ETSI TC ITS, based on the recent European ITS Communication Architecture, have paved the way towards vehicular cooperative systems. The ETSI proposal presents a more refined view of a communication stack that should be instantiated on Personal, Vehicle, Roadside and Central ITS Stations, where common OSI layers are surrounded by two planes for stack management and security. While the first one has been more or less exploited in terms of software lifecycle and interface management, above all in the CVIS project, the specification and implementation of the security plane is still a pending issue.

In current researches security is not taken into account in the communication stack development. Our proposal is a first attempt to integrate mobility services usually provided in vehicular scenarios with security mechanisms. This integration will result in a communication stack that also provides integrity and confidentiality to the transmitted traffic. Also, an extensible access control mechanism based on EAP [16] is part of this set of proposals that are distributed among the protocol layers of the ISO/ETSI ITS communication stack.

At link-layer level, Wi-Fi, 802.11p and 3G/UMTS communication technologies have been integrated with a network selection algorithm that can be parameterised according to preferences. At network-layer level, an IPv6 network mobility solution is provided to support the change of network attachment point when the communication stack is running on a vehicle. Also at network-layer level, secure IPv6 communications are achieved by means of standardized IETF protocols, such as Internet Protocol Security (IPsec) and Internet Key Exchange Version two (IKEv2). The final solution obtained gives a ready-to-use IPv6 communication stack provided with security mechanisms, which can be currently integrated in the ITS communication segment envisaged to be the first in being exploited: the vehicle to infrastructure (V2I) one or vice versa (I2V). Moreover, these capabilities have been validated by a suitable software platform and applications, and tested through performance tests obtained in real evaluations, whose main results are presented later.

It is worth noting that all the research has been developed using open source, and in particular, Linux distributions like Ubuntu and Busybox.

The outline of the rest of the chapter is as follows. Section 2 describes the concept of "Intelligent Transport System" (ITS), how the standard organizations like the International Organization for Standardization (ISO) and the European Telecommunications Standards Institute (ETSI) are working to develop the Reference ITS Communication Architecture, adding some arguments of how important is the IPv6 protocol for this architecture and ITS in general. An overview of the most common access technologies used in ITS scenarios can be found in Section 3, focusing on their suitability for ITS communications. Section 4 and 5 describe the ITS scenarios from the security and mobility point of view respectively. In Section 6 is discussed

http://dx.doi.org/10.5772/55285

the interoperability between security and mobility mechanisms. Finally, a set of evaluation results are presented in Section 7 followed by the conclusions in Section 8.

2. Intelligent Transport System (ITS)

Intelligent Transport System (ITS) applies advanced technologies of electronics, communications, computers, control and sensing and detecting in all kinds of transportation system in order to improve safety, efficiency and service, and traffic situation through transmitting real-time information.

Next subsection describes the Reference ITS Communication Architecture to be used in part of the FOTs scenarios. This architecture is mainly motivated by the works carried out by the International Organization for Standardization (ISO) and the European Telecommunications Standards Institute (ETSI) in the frame of Cooperative ITS. Its design and implementation has been done within the European IPv6 ITS Station Stack (ITSSv6) project, inside the 7TH Framework Program. As can be read in the rest of this section, it comprises a communication stack that can be instantiated into different roles that cover the key elements of a communication architecture based on the concept of ITS Station, from ETSI [5]. The communication stack implementation is totally based on IPv6 and offers capabilities especially focused on the network layer, also supporting a set of communication technologies and providing a common and simple IP view to facilities or final applications.

2.1. Reference ITS communication architecture

In an effort towards harmonization, the international ITS community agreed on the definition of a common ITS communication architecture suitable for a variety of communication scenarios (vehicle-based, roadside-based and Internet-based) through a diversity of access technologies (802.11p, 2G/3G, etc.) and for a variety of application types (road safety, traffic efficiency and comfort / infotainment) deployed in various continents or countries ruled by different policies.

This common communication architecture is known as the ITS station reference architecture and is specified by ISO [4] and ETSI [5]. This architecture is illustrated in Figure 1. Both ISO and ETSI architecture standards are based on the same terminology and tend to converge, although there are still remaining differences between the two. This lack of consistency shall disappear as standards are being currently revised.

It is divided in functional modules, each one with a defined and concrete functionality. The most important for our research are:

- IPv6 mobility management module: this module comprises mechanisms for maintaining IPv6 global addressing, Internet reachability, session connectivity and media-independent handovers (handover between different media) for in-vehicle networks. Nothing new, this module combines NEMO Basic Support [20] and Multiple Care-of Address Registration [21].

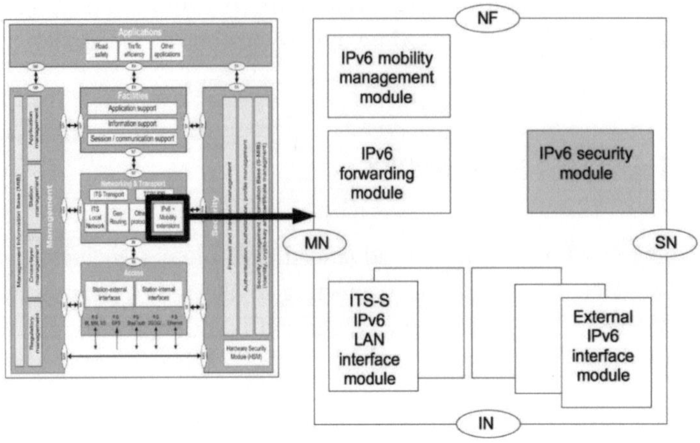

Figure 1. IPv6 functional modules in ISO 21210

It must only be present in ITS station IPv6 nodes performing functions to maintain Internet reachability and session continuity.

- IPv6 security module: the need for a module in charge of securing IPv6 communications is acknowledged in ISO 21210 [7], but the required features are not yet defined. In our case, it is assumed that this module comprises the functionality offered by IPsec [18], IKEv2 [15] and EAP [16].

One distinguishable feature of this architecture is the ability to use a variety of networking protocols in order to meet opposite design requirements i.e. fast time-critical communications for traffic safety versus more relaxed communication requirements for road efficiency and comfort / infotainment. However, for the majority of anticipated ITS applications and services, IPv6 is ideally suited. A particular emphasis was thus put on the use of IPv6 as the convergence layer ensuring the support of the diversity of access technologies, the diversity of applications and the diversity of communication scenarios. This resulted into the FP6 CVIS (Cooperative Vehicle-Infrastructure Systems) European Project taking a leadership on the specification and implementation of IPv6 Networking as defined by ISO [7], the FP6 SeVeCom (Secure Vehicular Communication) European Project investigating IPv6 security issues (IPv6 addresses based on pseudonyms), the FP7 GeoNet (IPv6 GeoNetworking) European Project specifying and implementing the concepts linking IPv6 networking and geographic addressing and routing, and the launch of IPv6-related work items at both ETSI TC ITS and ISO TC204. In addition, ETSI is going to provide test suites related to IPv6 and GeoNetworking.

2.2. ITS and IPv6

2.2.1. IPv6 basics

The first widely deployed protocol allowing packet-based communications between computers located in various networks was the Internet Protocol version 4 (IPv4) [8]. This protocol defines addresses of a fixed 32-bit length. This allows approximately 4 billion IP addresses to be used on the Internet. This figure appeared sufficient for the expected use of the protocol at that time. But the emergence of the commercial use of the Internet in the 90's decade led to an exponential use of IP addresses. To prevent the shortage of IP addresses, the IETF decided two measures: the specification of private IPv4 address spaces [9], to be used with Network Address Translation (NAT) and the design of a new version of the IP protocol: IP version 6 (IPv6).

The specification of this new protocol was finalized in 1998 [10]. This protocol defines addresses of a fixed 128-bit length. This allows a very large address space that is considered sufficient for most ambitious deployment scenarios (there would be enough addresses to identify every grain of sand on Earth). In addition to the address space, IPv6 defines new protocols to ease the management of the layer-3 protocol stack, such as Neighbour Discovery [11] that allows auto-configuration of IPv6 addresses.

While IPv6 is entering in its deployment phase, the depletion of the IPv4 address space is on going, despite the measures taken by the IETF. The global IPv4 address pool is exhausted since February 2011 and several regions such as Asia and Europe are facing shortage of IPv4 addresses. The exhaustion for the European region may happen in August 2012. After this date, Internet Service Providers (ISPs) and hosting services will not be able to get new IPv4 addresses. The deployment of IPv6 is therefore critical to ensure the future growth of the services of these stakeholders.

2.2.2. IPv6 for ITS

By the time ITS services requiring the use of the public IP addresses appear on the market, there will not be enough public IPv4 address available. The use of this version of IP scales to meet the addressing needs of a growing number of vehicles and connected devices, and provides the added functionality necessary in mobile environments. By relying on IPv6 in their ITS communication architectures, ISO, followed by ETSI, COMeSafety and the Car-to-Car Communication Consortium, have thus taken the right decision to guarantee sustainable deployment of Collaborative ITS.

Furthermore, IPv6 has the potential to decrease accident rates by enabling transmission of safety critical information. This chapter is not envisaged to demonstrate that this would be the case for real time applications, since the automotive industry and the SDOs(Standard Developing Organizations)at this time are not considering IPv6 for fast V2V communications. However, it is simple to note that not all data is time critical. There is no question that IPv6 could be a media-agnostic carrier of such non-time critical but safety essential information. FP7 GeoNet has demonstrated an example of the benefit of IPv6 for time critical application in the traffic hazard detection and notification scenario during its final demonstration [12]. Once the safety benefit of IPv6 is acknowledged, there are classical ways of calculating the economic impact of reducing road fatalities. E.g. the Safety Forum 2003 Summary Report estimated the cost of accidents at 160 billion euros. A 1% reduction would reduce these costs by 1.6 billion euros annually. And of course, this does not take into account the reduction of pain and suffering experienced by surviving family members and friends of accident victims that may not be adequately reflected in the method used to estimate the economic costs of traffic fatalities.

3. Access technologies

In order to provide access to a wide set of networks, the current distribution of the communication stack supports three key technologies in vehicular networking: 2G/3G, 802.11a/b/g (WiFi) and 802.11p. The communication stack is able to manage communication transceivers of many manufactures and it can exchange the data flow among them in a transparent way for the upper layers. All used frequencies have similar physical attributes, however their middle layer properties differ.

In the current work these interfaces have been provided by means of Laguna boxes from Commsignia (see Figure 2). The Laguna's firmware is ready to support these interfaces, providing an easy and unified way to configure them using the UCI (Unified Configuration Interface) from OpenWrt project [13].

Figure 2. Laguna box form Commsignia company

There are other wireless communication technologies considered for ITS communication like satellite, infra-red, WiMAX, microwave, millimetre wave; but they will not be evaluated during these tests.

3.1. GSM, 3G and UMTS

3G networks are wide area cellular telephone networks which have evolved to incorporate high-speed internet access. It has greater network capacity through improved spectrum efficiency. 3G technology supports around 144 Kbps, with high speed movement, i.e. in a vehicle, 384 Kbps locally, and up to 2Mbps for fixed stations, i.e. in a building. 3G technology uses CDMA, TDMA and FDMA, and the data are sent through packet switching.

http://dx.doi.org/10.5772/55285

3G has the best overall coverage or range of the three technologies and may serve as the easiest way of communication. Its range is 4-6 km at medium data throughput with the highest latency of the mentioned technologies. Response times can exceed 1 second.

The next generation technology is LTE (3GPP Long Term Evolution project) with speeds of 100/50 Mbit/s using orthogonal frequency-division multiplexing. The mobility support is also more robust, speeds up to 350-500 Km/h are supported depending on the frequency.

GSM also has the possibility to fall back to previous technologies below 3G (like EDGE) if no other types of services are present or are of low quality. This results in limited data throughput, but gives a higher technological reliability.

3.2. 802.11a/b/g/n

IEEE 802.11a/b/g/n protocols are supported by the Laguna's firmware and the compat-wireless open source project. Requirements of these standards are extremely and extensively complex, thus the implementations contain multi-level collaboration among device drivers, MAC layer modules, configuration and interface modules. Nevertheless, existing implementations support a wide range of wireless communication devices including upper layer functionalities. Summarizing, the existing solutions cover all the requirements against the IEEE 802.11a/b/g/n standards such as carrier frequencies, band width specifications, data rates, modulation schemes, authentication modes and security.

Operation and communication modes need to be supported according to the standards. One of these modes is the ad-hoc mode which is used for decentralized self-organizing connections between individual nodes. The implementation supports other, centralized modes: access point (ap) and station (sta). For these modes, authentication and encryption services are available as well, which are implemented by the Linux kernel and compat-drivers project [28].

High level configuration and management tools which are also integrated into the system prove a high level abstraction for device drivers and modules of the wireless framework. These user space applications provide a full scale management platform and most of them are integrated from the wireless-tools package e.g., iw, iwlib, etc. applications and libraries. Usually called as common WLAN (wireless local area networks), the a/b/g/n standards use the 2.4 and 5 GHz frequency bands with DSSS or OFDM modulation. It is a widely used technology with a range of 100-200 m. Response time calculation (using ad-hoc mode) supposes that several packets need to be transferred, before a link's status can be treated as set up.

An established connection may have a response time below 100ms either in ad-hoc or infra-structure mode.

In our test scenario, 802.11b is used for providing connectivity to the personal devices that use the car as point of attachment. It is capable of 11 Mbit/s maximum raw data throughput over the 2.4 GHz frequency band on 14 channels (split from 2.401 GHz to 2.495 GHz) using the original CSMA/CA protocol defined by the base standard. The common usage prefers the point-to-multipoint configuration. An access point communicates with one or more clients using an omni-directional antenna.

Common WiFi's latest amendment is 802.11n adding multiple-input multiple-output antennas used in both 2.4 and 5GHz to achieve net data rates up to 600 Mbit/s. The newest extension under development is 802.11ac, promising to better all parameters of the throughput.

3.3. 802.11p

The amendment modifies the 802.11 standard to add support for WLAN in a vehicular environment. It also proposes small modifications to the PHY and MAC layers in order to achieve a robust connection and a fast setup for moving vehicles. As 802.11a and g, the 802.11p is also based on the OFDM modulation method.

The 11p technology also differs by the fact that it is not as common as 802.11b or 3G and therefore has a limited hardware support, only a few companies offer transceivers capable of communicating with the required set of features and performance.

802.11p has the same range as WLAN, but uses a dedicated frequency (the licensed ITS band of 5.85-5.925 GHz) and an optimized set of protocols to reach a latency under all other listed technologies. Communication does not require a classic link to be set up and this results that high-priority traffic latency may be kept under tens of milliseconds at most. This attribute is essential for time-critical safety messages of ITS applications.

The implementation of the 802.11p feature is completely missing in the current open source world; therefore it cannot be imported from any known open source repository. Though there are several driver initiatives, a complete standard compliant version is yet to be found. The feature is newly developed to support the latest Linux kernel interfaces while focusing on the compliance of the latest communication standards.

4. IPv6 communication security

The main modules identified for providing security to the reference ITS communication architecture are IPsec and IKEv2. The latter is used to negotiate the security channels created by IPsec. A brief description of these technologies is included next.

4.1. IPsec

The IPsec software is present in Linux releases. The implementation supports both Encapsulated Security Payload (ESP) for both encryption and authentication and Authentication Header (AH) for authenticating the remote peer, which can be used together or separately to secure IPv4 or IPv6 traffic. The IPsec support is mandatory in IPv6 stack. The core implementation includes utilities for manual keying, while dynamic key management is implemented by other software components, using the IKE [14] and IKEv2 [15] protocols.

The main concept in IPsec is the Security Assiciation (SA), that establishes a secure channel to protect traffic. For selecting which traffic is going to be protected exists the concept of policy. This IPsec policies are stored in the Secure Policy Database (SPD). Every policy should be linked with a SA that determines the protocol (AH or ESP), cryptographic algorithm and keying material to be used to protect the traffic determined by the policy.

These SAs have to be established by hand by the networks administrators. As the keying material has to be refreshed with a determined frequency in order to ensure the traffic protection, administrators have to perform this task every time by hand, that becomes this in a bit tedious or even impossible task in large networks. For automate this task, IKE and IKEv2 protocols appeared, solving this scalability problem.

4.2. IKEv2

The Internet Key Exchange (IKE) protocol was designed in order to automate the IPsec Security Association (SA) establishment. The first version of IKE (IKEv1) [14] that was released in 1998, suffered some limitations and complexity, so that the IETF decided to propose a second version that was able to solve these limitations and simplified the protocol. The result was the IKEv2 protocol [15].

The IKEv2 protocol uses a non-reliable transport protocol (UDP using ports 500 and 4500) and it is performed between two parties: the initiator and the responder. As their names indicate, the initiator starts the IKEv2 communication, whereas the responder acts as server during the negotiation. The protocol is composed of a well-defined set of four main exchanges (request-response), namely: IKE_SA_INIT, IKE_AUTH, CREATE_CHILD_SA and INFORMATIONAL. Figure 3 shows the various IKEv2 message exchanges. These exchanges provides reliability to the IKEv2 protocol, since there is an expected and well defined response for each request.

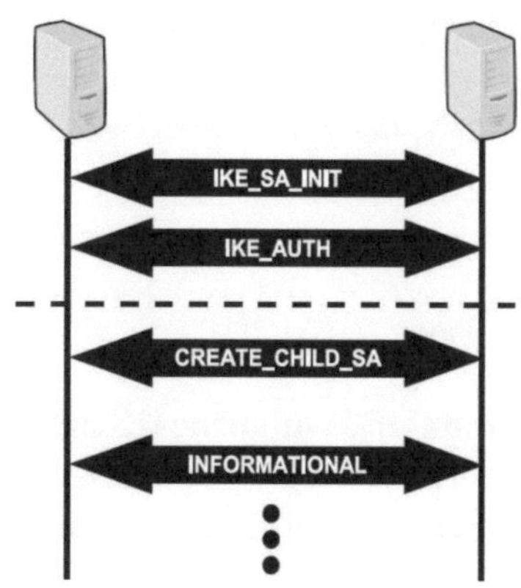

Figure 3. IKEv2 message Exchange.

The IKE_SA_INIT exchange establishes an SA at IKE level, named the IKE SA (IKE_SA), between the participant entities. This IKE_SA will protect all the following IKE exchanges. Once the IKE_SA is established, an IKE_AUTH exchange is performed in order to authenticate the parties and create the first IPsec SA (CHILD_SA) between them. These exchanges are denoted as initial exchanges and always must occur in this order.

There are two additional exchanges used for managing the SAs. The CREATE_CHILD_SA exchange allows creating additional SAs. The INFORMATIONAL exchange can be used for deleting SAs, notify events and manage configuration issues. When a SA expires a new one is created in order to replace the old one. This process is denoted as rekeying.

Additionally, IKEv2 introduces a good set of improvements with respect to IKEv1. For example, one of the main advantages of IKEv2 against IKEv1 is the inclusion of new features like NAT traversal, the transport of the Extensible Authentication Protocol (EAP) [16] for a flexible authentication mechanism and remote address configuration support.

4.3. EAP

Before establishing an IPSec based access control, the involved ITS station IPv6 nodes are required to authenticate each other. This mutual authentication is typically performed by means of the IKEv2 protocol [15], which in turn, relies on authentication mechanisms such as the Extensible Authentication Protocol [16], that allows the usage of a wide set interchangeable authentication methods.

5. IPv6 mobility in vehicular networks

Service continuity is a need when telematics services are offered in vehicles. Although it is possible to offer this feature at higher layers, maintaining communications at network level simplifies the implementation of facilities and applications in the ITS communication stack. The reference ITS station reference architecture described by ETSI and ISO includes, as part of its IPv6 support, Network Mobility Basic Support (NEMO) to accomplish the IPv6 communication continuity objective. The next parts of this section describe in more detail NEMO and some improvements that enhance the network connectivity of vehicles.

5.1. NEMO

5.1.1. IPv6 network mobility basic support (NEMO)

NEMO provides the necessary procedures within the ITS station networking & transport layer to allow an ITS station to maintain continuous IPv6 connectivity while changing its point of attachment to the network. The IPv6 mobility support module within the IPv6 protocol block ensures this. The IPv6 mobility support module comprises mechanisms for maintaining IPv6 global addressing, Internet reachability, session connectivity and media-independent handovers (handover between different access technologies) for in-vehicle networks. This module mostly combines Network Mobility Basic Support (NemoBS) [20]and Multiple Care-of Addresses Registration (MCoA) [21].

NemoBS is designed to maintain Internet connectivity between all the nodes in the vehicle and the infrastructure (network mobility support). This is performed without breaking the flows under transmission, and transparently to the nodes located behind the Mobile Router (MR), the mobile network nodes (MNNs), and the communication peers, also call "correspondent nodes" (CNs). This is handled by mobility management functions in the MR and a server known as the Home Agent (HA) located in an IPv6 subnet known to the MR as the home IPv6 link.

The key idea of NemoBS is that the IPv6 mobile network prefix (known as MNP) allocated to the MR is kept irrespective of the topological location of the MR while a binding between the MNP and the newly acquired temporary Care-of Address (CoA) configured on the external IPv6 egress connecting the MR to the Internet is recorded at the HA. This registration is performed by the MR at each subsequent point of attachment to an AR. In order to do this, the MR uses its global address known as the Home Address (HoA). This allows a node in the vehicle to remain reachable at the same IPv6 address as long as the address is not deprecated. The HA is now able to redirect all packets to the current location of the vehicle. MNNs attached to the MR do not need to configure a new IPv6 address nor do they need to perform any mobility support function to benefit from the Internet connectivity provided by the MR. This mobility support mechanism provided by NEMO is thus very easy to deploy, at a minimum cost.

The tunnel between the MR and the HA may be implemented as a virtual IPv6 interface pointing to a physical egress interface (external IPv6 interface) where packets would be encapsulated. The routing module as the physical external IPv6 interface would then treat such an IPv6 virtual interface. The same rules would thus be applied to the selection of the MR-HA.

The earlier Mobile IPv6 mobility support specification [22] provides Internet connectivity to a single moving IPv6 host only (IPv6 host mobility support). Mobile IPv6 is therefore inappropriate for the most advanced ITS use cases which usually consider more than one in-vehicle embedded CPU. Network mobility support using RFC 3963 also supports situations where there would be only a single IPv6 node deployed in the vehicle. Indeed, the ability to support an entire network of *n* nodes includes the ability to support a network of one node only. So just considering NEMO Basic Support, and not Mobile IPv6, makes the ITS station architecture much simpler.

The operation of NEMO Basic Support is illustrated in Figure 4.

For a better understanding of NEMO, the terminology is specified in [23] and the design goals behind NEMO Basic Support are described in [24]. These documents are normative documents about how to apply NEMO Basic Support to the ITS station architecture.

5.2. MCoA

The specifications given in [25] are extensions to Mobile IPv6 [22] and NEMO Basic Support [20] and allows a MR to register multiple Care-of Addresses (CoA) with its HA.

As a result of the notification of the tunnel set-up from the IP mobility management module to the ITS station management entity, the ITS station management entity should notify the IPv6 forwarding module with new forwarding table entries.

The operation of MCoA is illustrated in Figure 5.

The NEMO and MCoA extension of the Mobile IPv6 protocol was implemented as part of the NAUTILUS6 project. The project extended the UMIP (USAGI-patched Mobile IPv6 for Linux)

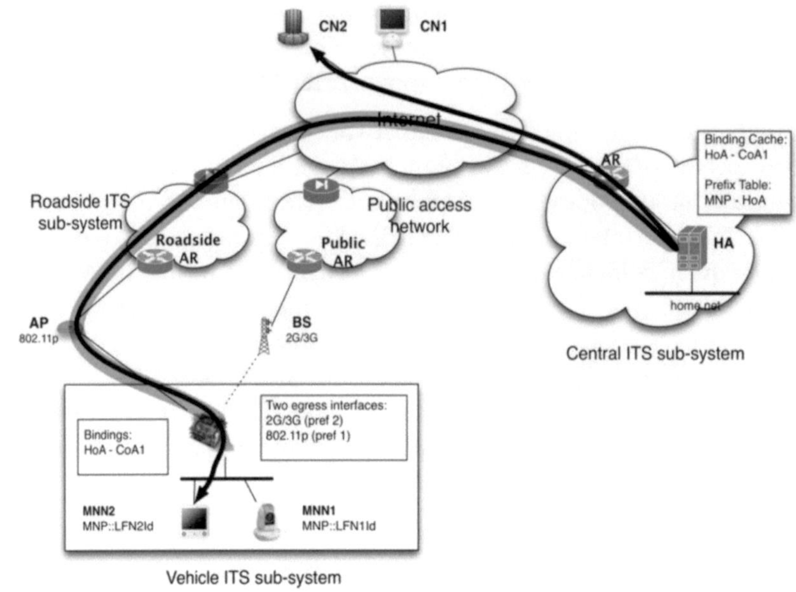

Figure 4. IPv6 session continuity with NEMO Basic Support.

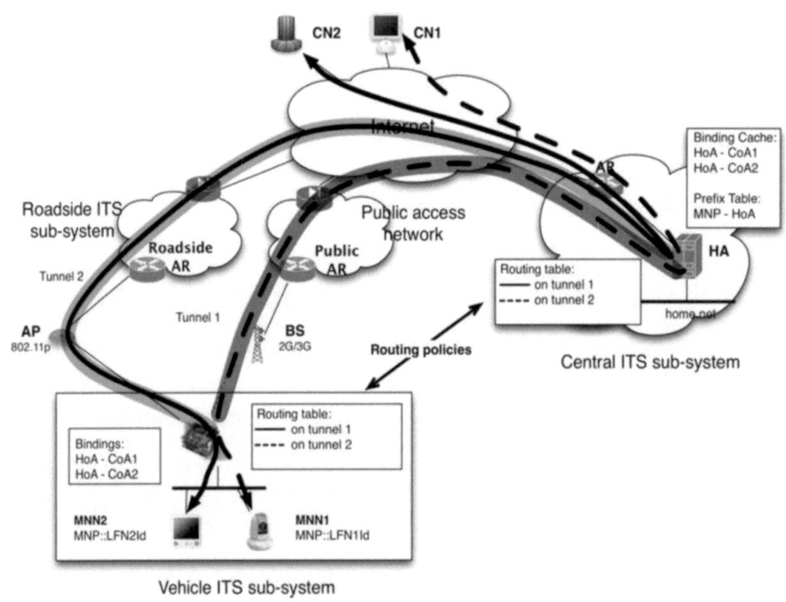

Figure 5. IPv6 mobile edge multihoming.

implementation with the NEMO, MCoA, DSMIPv6, HAHA, FMIPv6 protocols. After the project had ended an open source repository was opened on the UMIP.ORG website to continue the maintenance of the implementation with the help of the open source community. The MCoA protocol extension was not merged back into the code base because the standard was not finished at the time of the original implementation. To fully integrate the NEMO and MCoA functionality into the communication software stack presented in this chapter the existing implementation had to be reviewed in scope of standard compliance, scalability and conformance with other features such as IKEv2.

Advanced mobility support of IPv6 is two-fold in modern Linux environments. Packet transformation and decoding of protocol signalling implemented in the Linux kernel, while the rest of the protocol stack is implemented in a user space binary called mip6d. The latest open source snapshot of mip6d does not contain Multiple Care-Of Address support as the proposed implementation of the feature does not comply with the final standard definition.

The implementation of the MCoA protocol uses separate internal mechanisms for protocol signalling and data transmission. The description of internal procedures for handling protocols messages can be distinguished by the following properties:

- Signalling packets: Signalling messages such as Binding Updates (BU), Binding Acknowledgements (BA), Mobile Prefix Solicitation and Mobile Prefix Advertisement are built in the user space application and sent to the network stack via the socket interface. In case of BU/BA messages the XFRM framework is called which insert the Home Address Option and Routing Header Type 2 option headers after the IPv6 header. When IPsec is used, the installed XFRM policies demand that after the final packet structure is created, the payload of the packet is encrypted in ESP transport mode.

- Data packets: Data packets could originate from two sources: from the Mobile Router (MR) itself or from a Mobile Network Node (MNN). As MCoA allows the presence of multiple tunnels between the MR and the HA, all sharing the same Home Address (HoA) which makes exact routing decisions impossible. To properly route the packets, another identifier called BID is needed which selects the appropriate tunnel interface. The implementation design of the policy based routing mechanism is shown in Figure 6.

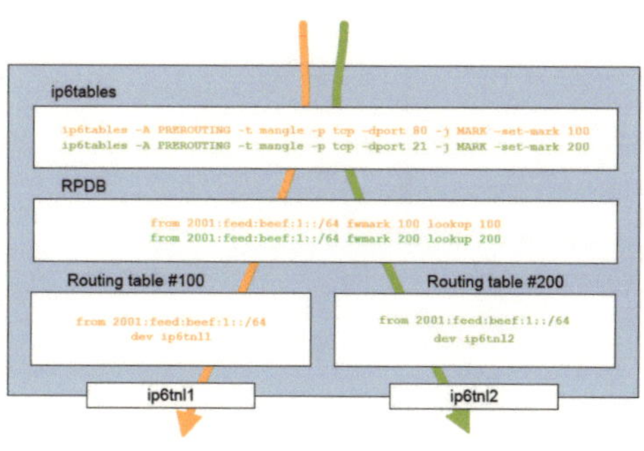

Figure 6. Policy routing used in MCoA.

Packet flows are marked by the Netfilter framework. Using the MARK target, all packets matching the rule are marked inside the kernel, so they can be later processed by the added mark value. Recent Linux implementations include a Routing Policy Database (RPDB), which allows the selection of routing tables based on a policy; in this case the routing table to use is determined by the packet mark value previously applied by the Netfilter framework. Each routing table contains default-route entries for different tunnel interfaces, hereby completing the policy routing method by sending the packet on the appropriate tunnel interface. IPsec secured data flows are routed the same way, however the encapsulation of data packets is done by the XFRM framework, which transmit the packets in ESP tunnel mode, making the above introduced ip6tnl IPv6-in-IPv6 kernel interfaces obsolete.

6. Integration of security, mobility and access control services

One of the main contribution in this chapter is the integration between the presented services of access control, security and mobility. There is no problem on using each service on its own, but when more than one is required to be applied at the same time, some interoperation issues have to be addressed first.

Figure 7 shows the considered scenario where a vehicle ITS station changes to a new point of attachment (Roadside ITS station). Following the standard mobility procedures, the vehicle's MR receives a RA message from the AR containing the network prefix (NP), which is used by the MR to configure the new MR's CoA (Care-of-Address).

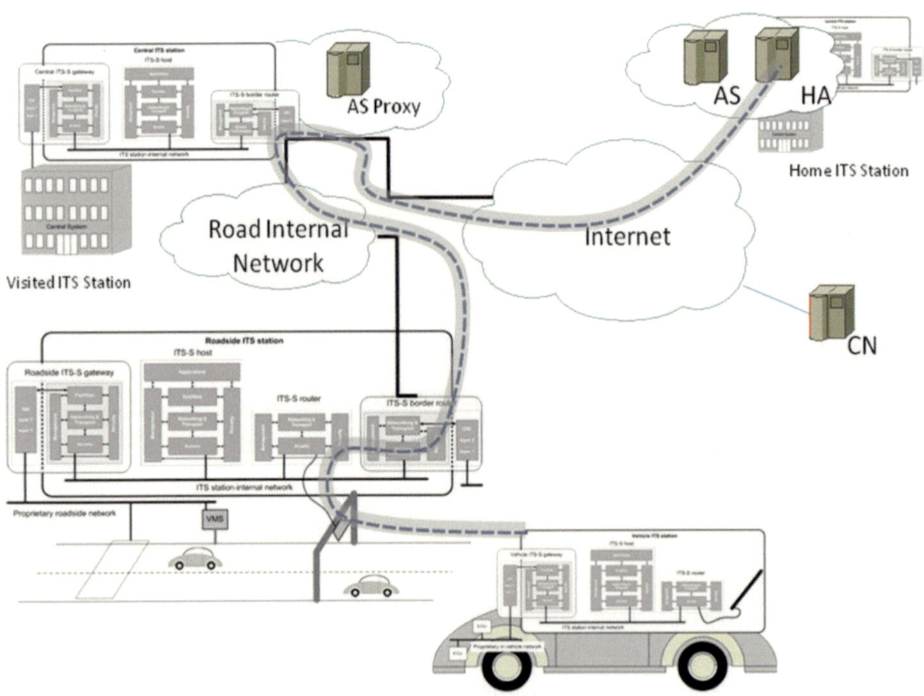

Figure 7. Authentication scheme for a vehicle ITS station accessing Internet through a roadside ITS station.

http://dx.doi.org/10.5772/55285

Before the use of the new configured CoA for IPv6 communications (e.g., to inform the home HA about the new vehicle's location), mobility signalling and data traffic are requested to be protected by means of IPsec. This protection has to be RFC4877 [26] compliant. This requires collaboration between security(IKEv2) and mobility(NEMO) services. First of all, IKEv2 has to manage the NEMO signalling traffic in a different and specific way. In addition, IKEv2 daemon needs to be aware of which of the available Care-of addresses is the one that has triggered the IKEv2 negotiation. For this reason, NEMO daemon has to notify this Care-of address to the IKEv2 daemon.

Before mobility negotiation could be performed, the MR must get authenticated against the AR to gain access to the network. This is performed by means of link-layer mechanisms like 802.1x [27]. Also, another authentication process has to be performed against the HA to gain mobility service. This authentication is performed jointly with the IPsec tunnel negotiation using IKEv2 protocol where EAP is used as authentication mechanism. While the MR acts as EAP peer, the HA implements the EAP authenticator functionality. The EAP authenticator may contact the home Authentication Server (AS) (acting as EAP server).

Once the EAP authentication is successfully completed, the IKEv2 protocol negotiates the parameters of the IPSec SA (keying material, algorithm, etc.). This IPSec SA is used to protect IPv6 packets transmitted between MR and HA through the ESP security protocol, thus satisfying the restriction on the HA for granting access to the mobility service to attached MRs.

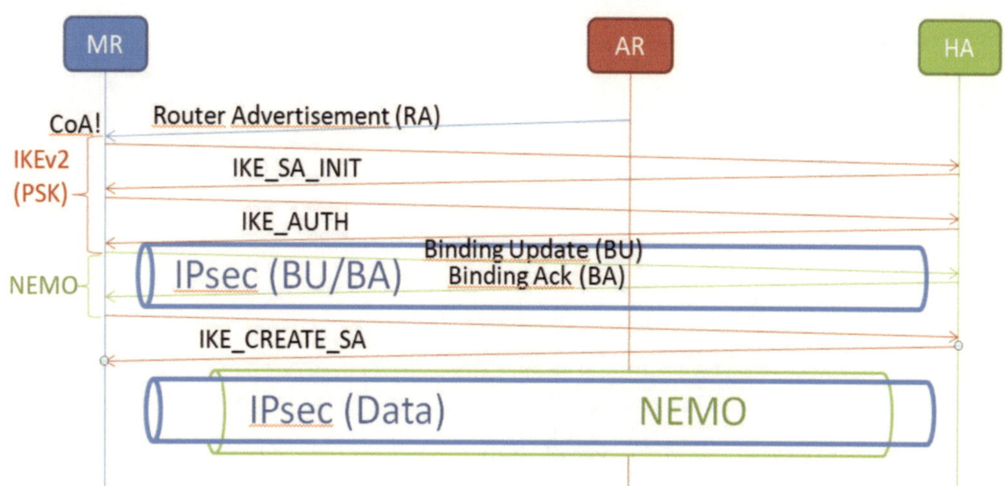

Figure 8. Sequence diagram of the IKEv2 and NEMO negotiation

In the Figure 8 can be seen that the negotiation is triggered by means of Router Advertisements (RA) presence, launching the IKEv2 negotiation to establish the IPsec tunnel that will protect the mobility control signalling, i.e., BU and BA messages. Once the mobility is also established, another IKEv2 negotiation is performed for the data traffic protection.

7. Performance evaluation of the secure network mobility solutions

The main purpose of the test is to evaluate the performance of the network when both mobility and security services are applied. Also, the performance of 802.11p wireless technology is measured, since it is aimed to be the next generation wireless technology designed for vehicular networks.

7.1. Description of the tests

In this scenario 3G and 802.11p can be used at the same time thanks to the MCoA capabilities of the mobility service.

As a transition mechanism for IPv4 support in a IPv6 network, an OpenVPN solution has been deployed, that creates IPv6 tunnels over IPv4 networks (3G networks), as it is depicted in Figure 9.

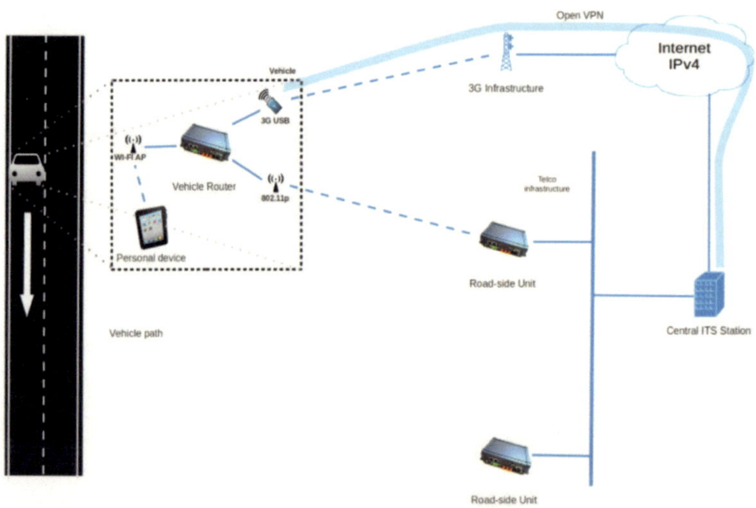

Figure 9. Test scenario using OpenVPN

In the test, the vehicle moves within the Espinardo Campus (University of Murcia) using the 3G connectivity and the 802.11p coverage to perform handoffs, as can be seen in Figure 10.

The next steps are carried out in the case of study considered for the tests:

1. The data flow start and the vehicle (MR) starts communicating through its home domain, through 802.11p at point A.
2. The vehicle (MR) moves towards a new (visited) domain, the one provided with the 3G infrastructure. An inter-domain and inter-technology handoff is necessary.
3. While the data connection is still maintained by the old data path through 802.11p, the vehicle (MR) connects to the 3G infrastructure, but it needs to gain network access

http://dx.doi.org/10.5772/55285

Figure 10. Test location

obtaining a new CoA. This is performed once the Router Advertisement message is received through the established OpenVPN tunnel.

4. The new MR CoA is registered in HA to change the data path used for both uplink and downlink communications.

5. The vehicle (MR) keeps moving and gains again 802.11p coverage. 802.11p is preferred and the MR changes its point of attachment. A new intra-domain and intra-technology handoff is necessary.

6. The vehicle (MR) keeps moving until point B and then stops.

The next metrics have been considered in the evaluations:

1. Bandwidth, measured in Mbps. It has been evaluated with a TCP flow maintained at the maximum allowable speed from the personal device to a correspondent node (CN) connected within the IPv6 UMU network.

2. Packet Delivery Ratio (PDR), measured in percentage of packets lost. It has been evaluated with a UDP flow in the downlink direction at 100 Kbps, 1Mbps and 5 Mbps, from the CN to the personal device.

3. Round-trip delay time (RTT), measured in ms. It has been evaluated with ICMPv6 traffic generated from the personal device to the CN. ICMPv6 Echo Request messages have been generated at a 1 Hz rate.

UDP and TCP traffics have been generated with the *iperf* utility, while the ICMPv6 traffic has been obtained from the common *ping6* Unix tool.

7.2. Results

The previous test plan has been executed enabling IPsec and IKEv2. Each of these round of tests considers one TCP trial, two UDP trials (100 Kbps and 1 Mbps), and one ICMP trial.

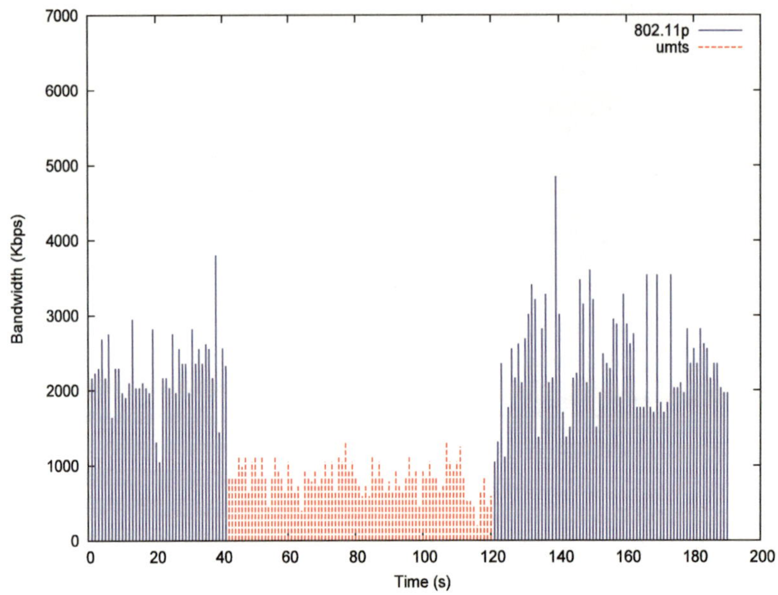

Figure 11. TCP bandwidth results

The bandwidth results obtained in the TCP tests are showed in Figure 11. As can be seen, the slow-start algorithm of TCP tries to adapt to the wireless medium in the whole test. This effect is even worse here if it is considered that tests have been carried out with a moving vehicle. Initially the vehicle is connected using 802.11p technology. The first handoff from 802.11p to 3G occurs just after time 40 sec, and the second one, from 3G to 802.11p at time around 120 sec. It is evident that the data rate using 802.11p is higher than 3G.

Although the achievable bandwidth is potentially higher in the 802.11p stretch than in any other part of the circuit, as it has been showed in the TCP case, the amount of packet losses is higher than when the 3G link is used as can be seen in Figure 12 and Figure 13. This is explained by the implementation of the 802.11p stack, which is in a very initial state where there are no improvements like Doppler effect mitigation. Changing the bit rate of the UDP transmission, we can appreciate that the packets are lost at the same locations in the path. It is worth nothing that in the handoff from 802.11p to 3G exists a gap where no transmission is possible, as you can appreciate in Figure 12 and Figure 13. This is due to the 802.11p coverage is gradually lost until the point that the transmission is not possible. The interface selector mechanism spends some seconds to realise that this interface is no longer usable and switches to the other one. This fact does not happen in the handoff from 3G to 802.11p because the 3G interface is not lost at anytime, so the interface selector mechanism switches seamlessly to 802.11p when this preferred interface is usable again.

http://dx.doi.org/10.5772/55285

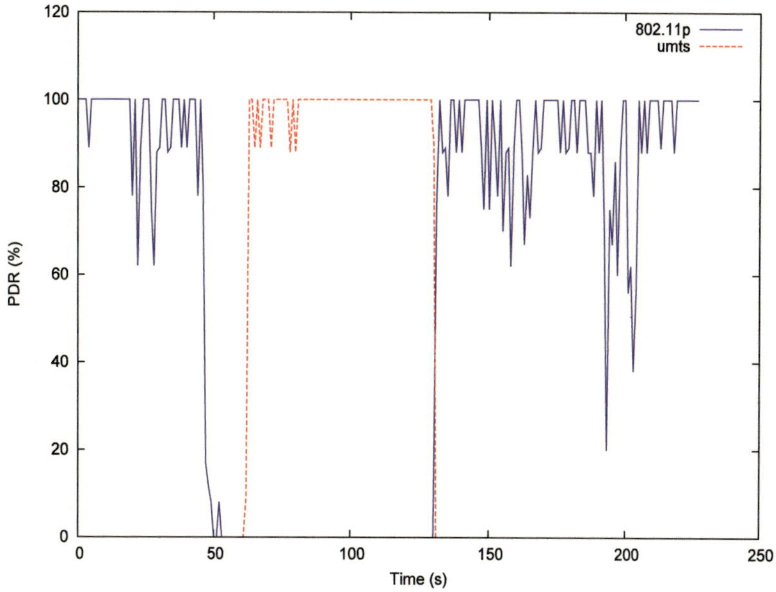

Figure 12. Packet delivery ratio at 100 Kbps

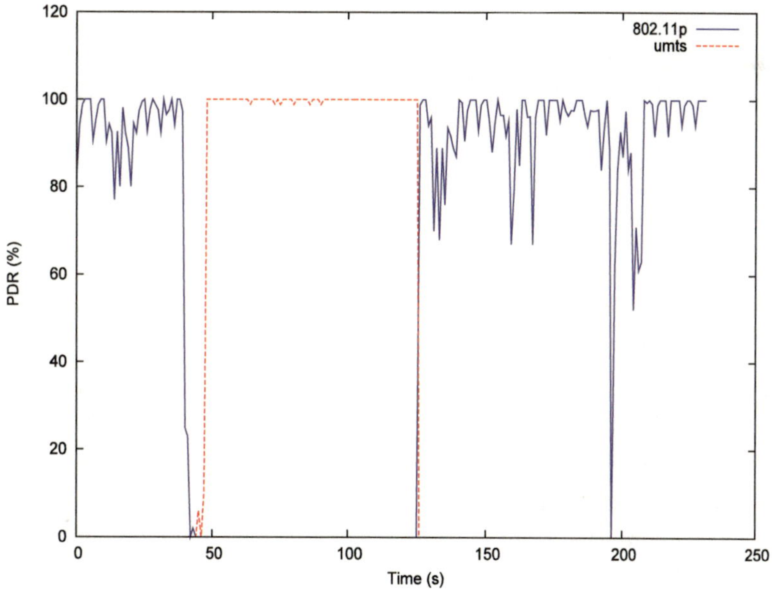

Figure 13. Packet delivery ratio at 1 Mbps

Finally, the network latency has been evaluated and results are given in Figure 14. These results have been obtained by generating ICMP traffic from the personal device inside the vehicle.

Regarding round-trip time (RTT), it is about twelve times better in 802.11p than in 3G. A low RTT is desirable for real-time applications, where the delay in packet delivery is crucial.

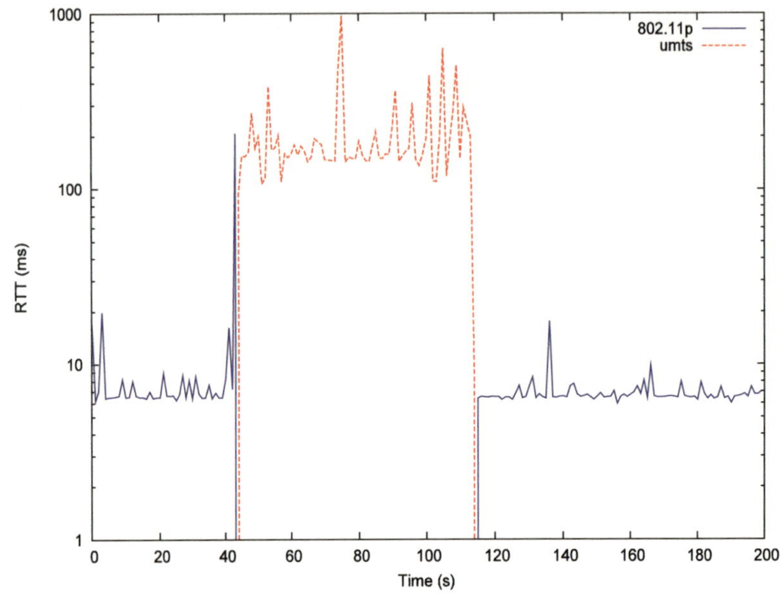

Figure 14. Latency evaluation with ICMPv6 traffic

8. Conclusion

This work presents a networking stack that follows current ESO/ETSI trends towards a common ITS communication architecture. The stack has been defined and developed, and it supports several communication technologies that can be automatically selected to provide connectivity to the vehicle. A secure IPv6 network mobility solution is also presented, using NEMO and an IPsec/IKEv2 combination to secure control and data traffic.

Taking into account the results presented in this chapter, this secure IPv6 network mobility solution performs efficiently under real inter-technology handoffs, the most difficult to accomplish. The communication stack operates correctly, maintaining the in-vehicle network connectivity in all tests and showing performance results that enable the communication stack to be used in many vehicular services. Unless high-quality multimedia transmissions are required, the bandwidth results indicate that the data rate required by most of the traffic efficiency and comfort services can be covered, and, according to latency tests, even non-critical security services, which are not highly dependent on real-time response, could be implemented, such as emergency assistance, variable traffic signalling or kamikaze warning.

http://dx.doi.org/10.5772/55285

Acknowledgements

This work has been sponsored by the European Seventh Framework Program, through the ITSSv6 Project (contract 270519), FOTsis (contract 270447); the Ministry of Science and Innovation, through the Walkie-Talkie (TIN2011-27543-C03) project; and the Seneca Foundation, by means of the GERM program (04552/GERM/06).

Author details

Pedro Javier Fernández Ruiz, Fernando Bernal Hidalgo, José Santa Lozano and Antonio F. Skarmeta*

*Address all correspondence to: skarmeta@um.es

Department of Information and Communication Engineering, University of Murcia, Spain

References

[1] Weib, C. V2x communication in europe: From research projects towards standardization and field testing of vehicle communication technology," Computer Networks, (2011). Deploying vehicle-2-x communication. Available: http://www.sciencedirect.com/science/article/pii/S1389128611001198, 55(14), 3103-3119.

[2] Festag, A, Le, L, & Goleva, M. Field operational tests for cooperative systems: a tussle between research, standardization and deployment," in Proceedings of the Eighth ACM international workshop on Vehicular inter-networking, ser. VANET'11. New York, NY, USA: ACM, (2011). Available: http://doi.acm.org/10.1145/2030698.2030710, 73-78.

[3] ISOIntelligent transport systems- Cooperative Systems- Terms, Definitions and Guidelines for Standards Documents- Part 1, April (2012). ISO/NP 17465:2012(E).

[4] ISOIntelligent transport systems- Communications Access for Land Mobiles (CALM)- Architecture, April (2010). ISO 21217:2010(E).

[5] Intelligent Transport Systems (ITS); Communications ArchitectureSeptember (2010). ETSI EN 302 665 , 1

[6] ISOIntelligent transport systems- Communications Access for Land Mobiles (CALM)- CALM using 2G Cellular Systems, April (2008). ISO/IS 21212:2008.

[7] ISOIntelligent transport systems- Communications Access for Land Mobiles (CALM)- IPv6 Networking, January (2011). ISO 21210:2011(E)

[8] Internet Engineering Task ForceRFC 791 Internet Protocol- DARPA Inernet Programm, Protocol Specification, September (1981).

[9] Rekhter, Y, Moskowitz, B, Karrenberg, D, De Groot, G. J, & Lear, E. Address allocation for private internets. RFC 1918 (Best Current Practice), February (1996).

[10] Deering, S, & Hinden, R. Internet Protocol, Version 6 (IPv6) Specification. RFC 2460 (Draft Standard), December (1998). Updated by RFCs 5095, 5722, 5871.

[11] Narten, T, Nordmark, E, & Simpson, W. Neighbor discovery for ip version 6 (ipv6). RFC 2461 (Proposed Standard), (1998).

[12] GeoNetD7.1 GeoNet Experimentation Results. Public deliverable, June (2010).

[13] OpenWRT: Linux distribution for embedded deviceshttps://openwrt.org (accessed 27 July (2012).

[14] Harkins, D, & Carrel, D. The Internet Key Exchange (IKE). RFC 2409 (Standards Track),November (1998).

[15] Kauffman, C. Internet Key Exchange (IKEv2) Protocol. IETF RFC 4306, Dec. (2005).

[16] Aboba, B, Blunk, L, Vollbrecht, J, Carlson, J, & Levkowetz, H. Extensible AuthenticationProtocol (EAP). RFC3748, June (2004).

[17] Arkko, J, Kempf, J, Zill, B, & Nikander, P. Secure neighbor discovery (send). RFC 3971 (Proposed Standard), March (2005).

[18] Kent, S, & Seo, K. Security architecture for the internet protocol. RFC 4301, (2005).

[19] Jari Arkko, Vijay Devarapalli, and Francis Dupont Using IPsec to protect mobile ipv6 signalling between mobile nodes and home agents. RFC, June (2004).

[20] Devarapalli, V, Wakikawa, R, Petrescu, A, & Thubert, P. Network Mobility (NEMO) Basic Support Protocol. RFC 3963 (Proposed Standard), January (2005).

[21] Wakikawa, R, Devarapalli, V, Tsirtsis, G, Ernst, T, & Nagami, K. Multiple Care-of Addresses Registration. RFC 5648 (Proposed Standard), October (2009).

[22] Johnson, D, Perkins, C, & Arkko, J. Mobility Support in IPv6. RFC 3775 (Proposed Standard),June (2004). Obsoleted by RFC 6275.

[23] Ernst, T, & Lach, H-Y. Network Mobility Support Terminology. RFC 4885 (Informational), July (2007).

[24] Ernst, T. Network Mobility Support Goals and Requirements. RFC 4886 (Informational), July (2007).

[25] Wakikawa, R, Devarapalli, V, Tsirtsis, G, Ernst, T, & Nagami, K. Multiple Care-of Addresses Registration. RFC 5648 (Proposed Standard), October (2009).

[26] Devarapalli, V, & Dupont, F. Mobile IPv6 Operation with IKEv2 and the Revised IP-sec Architecture.RFC 4877 (Standards Track), (2007).

[27] IEEE 802x- Port Based Network Access Control- http://www.ieee802.org/1/x.htmlac-cesed by 30th July (2012).

[28] Compat-drivers project- http://linuxwirelessorg (accesed by 27th November (2012).

Section 3

Access in Vehicular Networks

Chapter 5

Access Control and Handover Strategy for Multiple Access Points Cooperative Communication in Vehicle Environments

Xiaodong Xu

Additional information is available at the end of the chapter

http://dx.doi.org/10.5772/55401

1. Introduction

Current communication technologies applied in vehicle environments meet a lot of challenges, such as larger capacity requirements, higher velocity scenarios and lower latencies. Among these challenges, higher velocity scenarios addressed most of focuses in recent years. With the world-wide rapid constructions and deployments of high-speed train transportation system, the mobile communication support for high speed vehicle environments need further improvement.

In Dec. 2004, 3GPP (3rd Generation Partner Project) launched LTE (Long Term Evolution) standard work, which improved the mobile speed support that should be up to 350km/h and even 500km/h in some frequency bands [1]. The same requirements are also included in LTE Advanced system [2]. The higher speed vehicle environments will introduce larger Doppler frequency offset which need to be coped with in physical layer techniques. Furthermore, frequent handover and access process will also occur in the high speed vehicle communication scenarios, which need the evolution of the Media Access Control (MAC) and Radio Resource Management (RRM) layer techniques.

With the research and development of mobile/vehicle communication systems, a lot of advanced physical layer technologies show their merits and are applied in next generation mobile telecommunication systems. Among these techniques, the multi-antenna techniques, such as MIMO and OFDM, show their merits in improving system capacity and coverage area [3]. In LTE Advanced system, cooperative communication is introduced with CoMP (Coordinated Multi-Point) technique [4]. CoMP implies dynamic coordination communication among multiple separated transmission points, which can improve the received signal quality and cell edge user performances.

Cooperative communication needs more than one access point to accomplish the communication process. And with the developments of communication infrastructure construction, actually we have more densely deployments of base stations or access points in next generation system now. The optical fibre enlarges the coverage and also increases the numbers of RF (radio frequency) head. For the high speed train transportation system, there will also be lots of access points deployed along the railway.

Multiple access points communication environments increase the potentiality of larger communication capacity and higher speed vehicle support. But the multiple access point communication environments also introduce the challenge for traditional access and handover strategies.

The traditional access control and handover algorithm cannot accommodate the features of multi access point communication scenarios, especially for users served by multiple distributed antennas. Therefore, this chapter will focus on the Access Control and Handover strategies for multiple access point cooperative communication in vehicle environments.

In the following parts, the research outcomes about Access Control [5,6], Slide Handover strategy [7] and corresponding performance evaluation results will be provided. Maximum Utility Principle will be proposed and applied in the Access Control and Slide Handover process.

In part 2, the MUPAC (Maximum Utility Principle Access Control) based Dijkstra's Shortest Path Algorithm for multiple access point vehicle environment will be introduced. In the accessing process, the shortest path in Dijkstra's Algorithm can be represented by the cost of accessing process, which is formed by utility function. Based on the multiple access point cooperative communication vehicle environment, the MUPAC algorithm is described in details with the utility function, Maximum Utility Principle, flow chart of accessing process. Scheduling strategy enhanced version of MUPAC and corresponding performance evaluation verify the merits of MUPAC algorithm in improving system capacity, accessing success probability and efficiency of system resources usage.

Part 3 describes the Maximum Utility Principle Slide Handover strategy for multiple access point vehicle environments. Slide Handover strategy is illustrated and its merits will be presented. Slide window is applied in the Slide Handover process, which makes users always stay in cell centre and eliminates cell-edge effect. Maximum Utility Principle is also applied in Slide Handover strategy, which can effectively solve this problem. The Utility Function in the Slide Handover and steps for handover are described and system-level performance evaluations are also provided.

Finally, there comes the summary for this chapter.

2. Maximum utility principle access control strategy for vehicle environments

In [5], we brought out a MUPAC method for multi-antenna cellular architecture with application in Group Cell architecture [8] as an example. MUPAC method can maximize the usage

http://dx.doi.org/10.5772/55401

of limited system resources with guaranteeing access users' QoS requirements. Furthermore, through this method, the interference caused by access users can also be mitigated maximally and the accessing success probability can also be improved.

With MUPAC method, when the system is under heavy-load situation, MUPAC method can fully show its merits in improving resource efficiency. However if the system is relatively light loaded and the capacity is enough for most of users, MUPAC method may not fully use system capability to serve users with its best, because MUPAC cares more on the minimum QoS requirements for resource utility rather than on user better performance. So, scheduling can be used in combination with MUPAC method to solve this problem and improve user service experience after access success ratio improvements.

This part will present two improvement based on MUPAC with scheduling. One is Throughput Targeted MUPAC (TT-MUPAC) and another is Throughput and Fairness Targeted MUPAC (TFT-MUPAC) algorithm [6].

2.1. Maximum utility principle access control

Taking Group Cell architecture [8] as the typical application scenario of multiple access point communication vehicle environments, users in the system are served by more than one antenna/access point. The access control method in this situation needs to solve the problem of how to choose multiple antennas to form the serving Group Cell and allocate appropriate resources to users. The size of Group Cell can be adjustable for users by their QoS requirements. Therefore, we can add antenna with maximum utility to user's current serving Group Cell step by step and fulfil the users' QoS requirements. This solution can maximum the usage of limited system resources with guaranteeing access users' QoS requirements. Furthermore, the interference caused by new users can also be mitigated maximally and the accessing success probability can also be improved.

The steps of adding antennas with maximum utility can be accomplished based on Dijkstra's Shortest Path Algorithm [9] in Graph Theory. Based on Dijkstra's Shortest Path Algorithm, when there are new users initiate their access attempts in Group Cell architecture, the shortest path in the Dijkstra's Algorithm can be replaced by the minimal cost of accessing process. The cost of accessing process includes the interference to other users and system resources needed (antennas, channels or other resources). Furthermore, the cost can be represented by the utility functions, including the gain for the access user and deterioration to other users. Therefore, the seeking for shortest path in Dijkstra's Algorithm can be transferred to seek the antennas or resources with maximum utility. The Maximum Utility Principle can improve the system capacity and load ability. By the Dijkstra's Shortest Path Algorithm and the Maximum Utility Principle, the user accessing in multi-antenna distributed Group Cell can be effectively accomplished.

In MUPAC method, the utility function is constructed by considering current system load, resource employment and so on. The utility function has two objectives. One is used to add antennas to current serving Group Cell with Maximum Utility Principle and the other is to select system resources allocated to the access users with Maximum Utility Principle.

The utility function has two aspects, including the gain of new antenna added in current serving Group Cell and the deterioration for the other users existed in the system.

The utility function is shown as (1).

$$\begin{aligned} U(i, ..., j, k) &= \zeta_{Ck}[G_C(i, ..., j, k) - I_C(i, ..., j, k)] \\ &+ \beta(1-\zeta_{Ck}) \max_{M \neq C}\{\zeta_{Mi} \cdot ...\zeta_{Mj} \cdot \zeta_{Mk}[G_M(i, ..., j, k) - I_M(i, ..., j, k)]\} \end{aligned} \quad (1)$$

where $U(i, ..., j, k)$ denotes the utility of adding antennas k to current serving Group Cell formed by antennas $i, ..., j$. C and M denote the resources and C is the current resource used by serving Group Cell. ζ_{Ck} is an indicator function, which indicates the occupying information of resource C in AE (antenna element) k.

$$\zeta_{Ck} = \begin{cases} 0, \text{ Resource C occupied in AE k} \\ 1, \text{ Resource C available in AE k} \end{cases} \quad (2)$$

where, $G_C(i, ..., j, k)$ denotes the gain achieved by adding antennas k to current Group Cell with resource C. $I_C(i, ..., j, k)$ denotes the interference to other users by adding antennas k to current serving Group Cell with C. β is a constant between 0 and 1 to introducing the penalty for coordinating current resource C and different resource (resource C') for the new serving Group Cell. β can be set according to the current system load condition. The choice of C' replacing C can also be achieved by Maximum Utility Principle with the utility function, which is:

$$C' = \arg\max_{M \neq C}\{\zeta_{Mi} \cdot ...\zeta_{Mj} \cdot \zeta_{Mk}[G_M(i, ..., j, k) - I_M(i, ..., j, k)]\} \quad (3)$$

Specifically, when choosing the first antennas to form the serving Group Cell, the utility for choosing the first antennas can be written as:

$$U(k) = \max_{M}\{\zeta_{Mk}[G_M(k) - I_M(k)]\} \quad (4)$$

when ζ_{Mk}, $G_C(i, \ldots, j, k)$ and $I_C(i, \ldots, j, k)$ in (1) do not exist before serving Group Cell is formed. Correspondingly, the method for choosing system resource for the new Group Cell by Maximum Utility Principle can be written as:

$$C = \arg\max_{M}\{\zeta_{Mk}[G_M(k) - I_M(k)]\} \quad (5)$$

Considering the actual mobile systems, the gain and interference in utility function are usually represented by SINR. Therefore, (1) can be revised to:

http://dx.doi.org/10.5772/55401

$$
\begin{aligned}
&U(i, ..., j, k) \\
&= \zeta_{Ck}[\frac{\lg_{k,i}}{\sum_{n \neq i, j...,k} (1-\zeta_{Cn}) \lg_{n,i}} - \sum_{n \neq i,...,j,k} (1-\zeta_{Cn}) \frac{\lg_{n,n}}{\lg_{k,n}}] \\
&+ \beta(1-\zeta_{Ck}) \underset{M \neq C}{\operatorname{argmax}} \left\{ \begin{array}{l} \zeta_{Mi}\zeta_{Mj}...\zeta_{Mk} \\ [\frac{\lg_{k,i}}{\sum_{n \neq i,...,j,k} (1-\zeta_{Mn}) \lg_{n,i}} - \sum_{n \neq i,...,j,k} (1-\zeta_{Mn}) \frac{\lg_{n,n}}{\lg_{k,n}}] \end{array} \right\}
\end{aligned} \tag{6}
$$

And (4) can also be revised to:

$$
U(k) = \underset{C}{\operatorname{argmax}} \left\{ \zeta_{Ck} \left[\frac{\lg_{k,k}}{\sum_{n \neq k} (1-\zeta_{Cn}) \lg_{n,k}} - \sum_{n \neq k} (1-\zeta_{Cn}) \frac{\lg_{n,n}}{\lg_{k,n}} \right] \right\} \tag{7}
$$

where $\lg_{n,k}$ denotes the path gain Between the antenna n and the access user who is currently served by the antenna k. The power for each user in (6) and (7) are equally allocated.

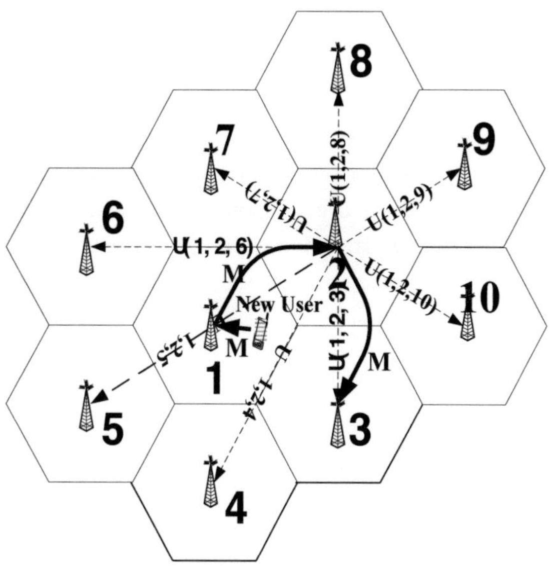

Figure 1. MUAC method in Group Cell

The application example of MUPAC in Group Cell architecture is shown as Fig. 1. The detailed implementation steps of MUPAC are shown as follows.

1. Access user initiates access attempt.
2. AP obtains the users' receiving pilot strength of each antenna.

3. Based on the information in step 2), AP calculates the utility of each antenna and available resource by utility function and chooses the first antenna and resource with the Maximum Utility Principle to form the serving Group Cell. If all the antennas detected by access user have no resource available, the access user will be transferred to the accessing waiting list. In Figure 1, the access user selects AE 1 as the first serving antenna with resource unit by Maximum Utility Principle.

4. AP obtains the users' receiving SINR of serving Group and compares it with user's QoS requirement. If current serving Group can provide adequate QoS to the user, the access process accomplishes successfully. Vice versa, the access user need more antennas added to the current serving Group. In Figure 1, the access user needs more antennas to get its desired QoS. So, AE 2 is chosen to be added in current serving Group Cell.

5. AP obtains the users' receiving SINR of antennas excluding current serving Group and chooses the antenna with maximum utility to add it to the serving Group Cell. This step needs to guarantee the new antenna and current serving Group Cell to use the same resource. The utility function includes the penalty of resource changing. Then, goes to step 4). In Figure 1, AE 3 is added and the serving Group Cell of AE 1, 2 and 3 has enough quality to serve the user with resource unit M.

2.2. Maximum utility principle access control with scheduling

When we are choosing the algorithms for access control, we always care about the quality of services, as well as the efficiency of resource assignment which is associated with the system capacity. Ensuring the QoS (quality of service) of access users' communications, MUPAC method gives the least sources to users to reach a minimum acceptable QoS. Considering the variable mobile communication environments and multi-user diversity, also the service experience of users, it will be helpful to implement scheduling into the process of access control.

2.2.1. Throughput Targeted-MUPAC (TT-MUPAC)

In order to make better use of the resources and reach higher system throughput, we should consider using scheduling in access control to adapt to different environments and make full use of the resources. When the system load is light, MUPAC is not good enough, especially in the condition of dealing with data services. If we make full use of system resources and increase the system throughput, it would be beneficial to either the users or the system. Throughput Targeted-Maximum Utility Principle Access Control (TT-MUPAC) brings out a good consideration on this point.

TT-MUPAC strategy gives different resources to different users in access control which depends on the system conditions. If the system is heavy-loaded with many services required, it gives the user the least resource to reach the required QoS. On the other hand, if the system is relatively light-loaded and there are many resources available, the access users will get most resource to improve system throughput.

In TT-MUPAC strategy, MAX C/I scheduling algorithm is employed. The key point of combination of MAX C/I and MUPAC is to give some users more resource to get multi-user diversity in the system. In this way, we can improve the system throughput obviously.

2.2.2. Throughput and Fairness Targeted-MUPAC (TFT-MUPAC)

TT-MUPAC brings some advantages on system throughput, but when it comes to user fairness, the performance is decreased. Throughput and fairness are both important in access control strategy. So we should make some improvements on MUPAC and TT-MUPAC methods to reach a better performance on throughput and fairness. Throughput and Fairness Targeted-Maximum Utility Principle Access Control (TFT-MUPAC) method is proposed to achieve a balance between fairness and system throughput.

In the TFT-MUPAC strategy, the utility function should consider both system throughput and user fairness. In order to include the consideration of fairness into the access control strategy, we add a fairness factor into the utility function to present the improvement based on formula (1).

$$\begin{aligned} U'(i, \ldots, j, k) &= F \cdot G(i, \ldots, j, k) = \\ &\left(\frac{Raverage}{Rgenerated}\right)^{\gamma} \cdot \zeta_{Ck}[G_C(i, \ldots, j, k) - I_C(i, \ldots, j, k)] \\ &+ \beta(1-\zeta_{Ck})\max_{M \neq C}\{\zeta_{Mi} \cdot \ldots \zeta_{Mj} \cdot \zeta_{Mk}[G_M(i, \ldots, j, k) - I_M(i, \ldots, j, k)]\} \end{aligned} \tag{8}$$

F denotes the fairness of the service quality, which is,

$$F = \left(\frac{Raverage}{Rgenerated}\right)^{\gamma} \tag{9}$$

$G(i, \ldots, j, k)$ denotes the gain achieved by adding antenna k to current serving group. $R_{generated}$ is the service quality user can get with the target antennas when he is accessed. $R_{average}$ is the average service quality of the users already in the system. γ is the factor of fairness.

In the TFT-MUPAC strategy, antennas are allocated to receive a fairer QoS. At the same time the system throughput is also considered. Proportional fairness scheduling method is employed. In this way, system carries out a good performance on both system throughput and fairness.

2.3. Performance evaluation

For the performance evaluation and analyses, MUPAC method is taken for performance comparing based on Group Cell architecture. MUPAC chooses antennas and allocates resources according to the Maximum Utility Principle. The access point number of Maximum Utility Principle Access Control method is limited to 4. TT-MUPAC and TFT-MUPAC employs scheduling with MAX C/I and Proportional Fairness algorithms. System-level simulation is adopted to evaluate these three access control methods by comparing the successfully accessed user numbers with different system load (total access user number generated), system throughput and fairness. The power allocation for these three algorithms is the same as fixed power allocation scheme. The simulation parameters and setting are shown in Table 1.

Parameters	Setting
Traditional inter-site distance	$500\sqrt{3}$m
Group Cell inter-antenna distance	500m
Carrier Frequency	5.3GHz
Path gain model	25log10(d)+35.8 [10]
Shadow fading deviation	5dB
Total bandwidth	20MHz
Effective bandwidth	17.27MHz
Number of useful sub-carriers	884
Sub-carrier spacing	19.5KHz

Table 1. Simulation parameters and setting

The simulation results are shown in Figure 2, Figure 3, Figure 4, Figure 5, Figure 6 and Figure 7.

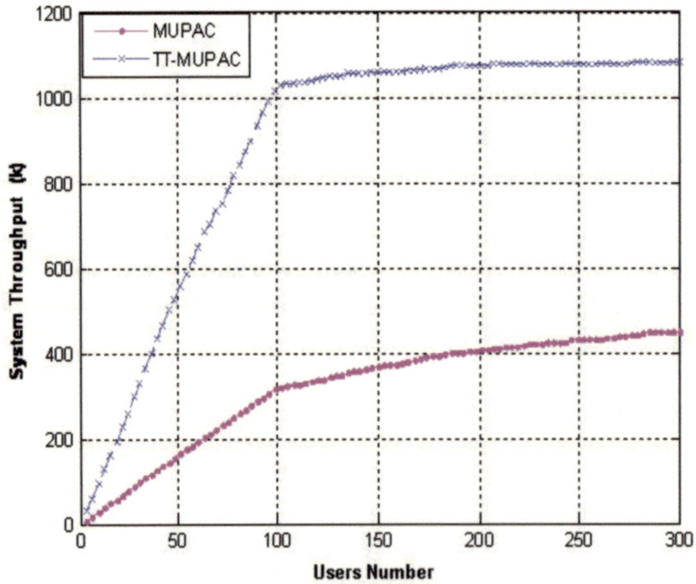

Figure 2. System Throughput of MUPAC vs. TT-MUPAC

Figure 2 shows system throughput of MUPAC and TT-MUPAC. TT-MUPAC has obvious throughput advantage over MUPAC scheme. The reason for this throughput gain mainly comes from multi-user diversity with MAX C/I scheduling. MUPAC only guarantees the minimum requirements of access users' QoS for maximum resource efficiency. By TT-MUPAC, scheduling is able to improve the throughput with light load.

Figure 3 shows access success rate of MUPAC and TT-MUPAC. MUPAC is better than TT-MUPAC, because TT-MUPAC use more resources for few users to get more throughputs. The relatively low efficiency of resource utility makes access users have less available resources and lows the access success rate.

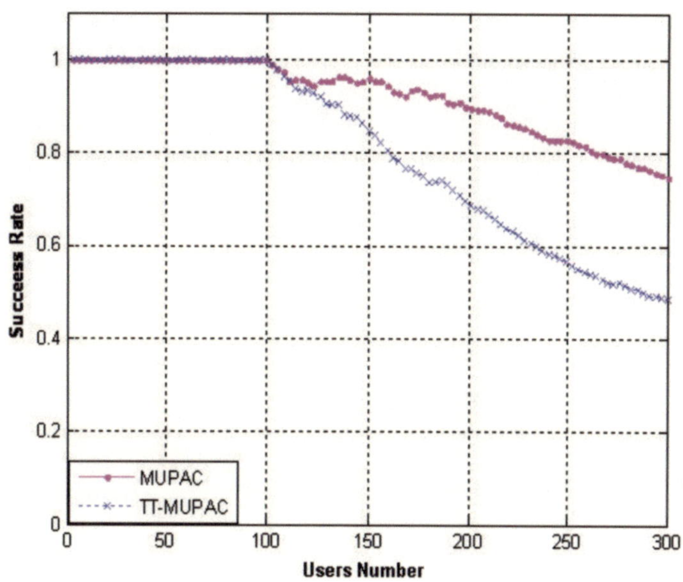

Figure 3. Access Success Rate of MUPAC vs. TT-MUPAC

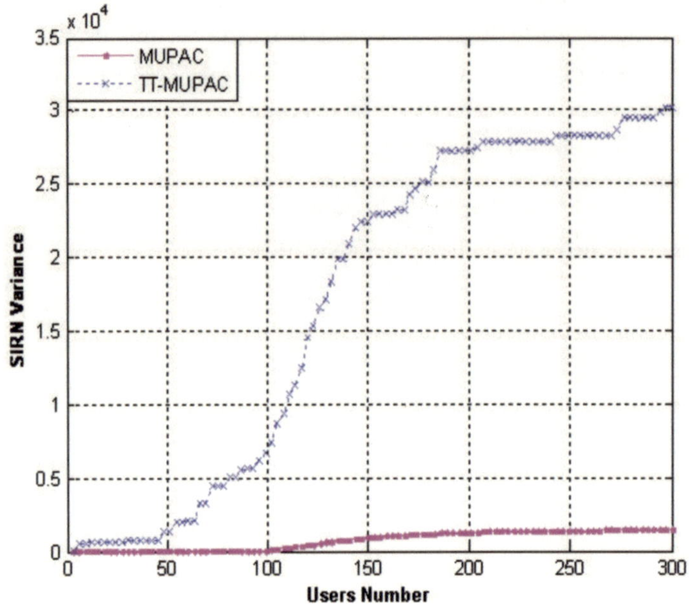

Figure 4. System Fairness of MUPAC vs. TT-MUPAC

Figure 4 shows the fairness of access users based on MUPAC and TT-MUPAC by SINR variance. From the simulation results, TT-MUPAC has worse fairness than MUPAC, which dues to the MAX C/I scheduling method.

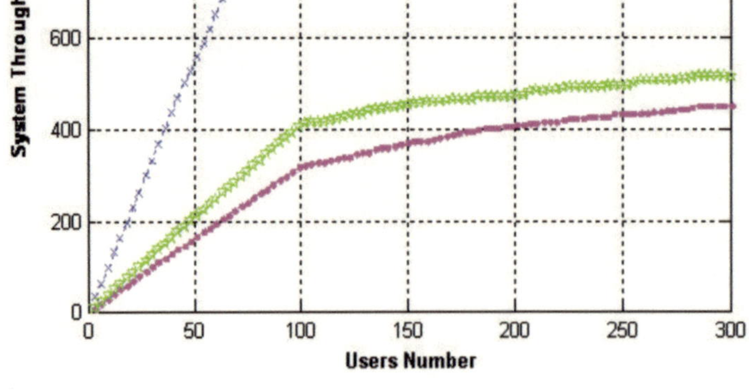

Figure 5. System Throughput of MUPAC, TT-MUPAC and TFT-MUPAC

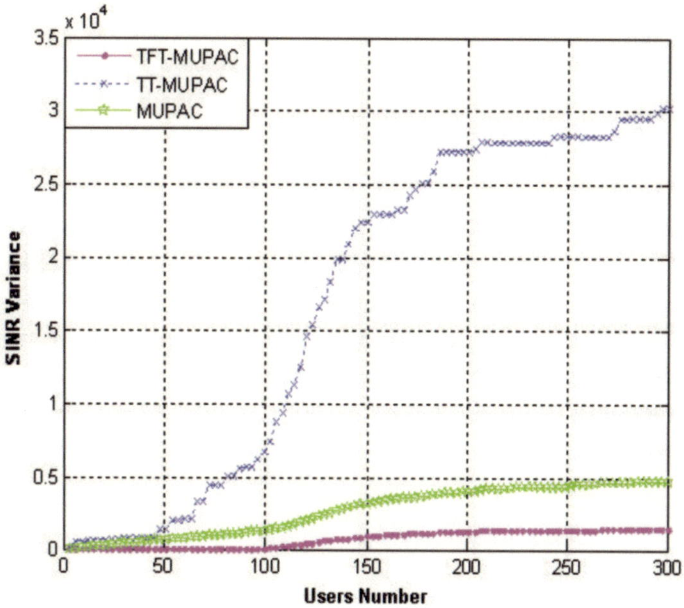

Figure 6. User Fairness of MUPAC, TT-MUPAC and TFT-MUPAC

Figure 5 and figure 6 show the performance of MUPAC, TT-MUPAC and TFT-MUPAC, including system throughput and user fairness. Figure 5 shows the throughput performance of MUPAC, TT-MUPAC and TFT-MUPAC. TT-MUPAC has the best performance and TFT-MUPAC has the worst performance with the features of scheduling methods. Figure 6 shows the user fairness of these three methods. TFT-MUPAC has better fairness performance than MUPAC and TT-MUPAC.

3. Maximum utility principle slide handover for vehicle environments

3.1. Slide handover strategy

Based on Group Cell architecture [8], Slide Handover strategy is proposed with multi-antenna slide windows [11]. Slide Handover makes user always in the centre of its corresponding serving Group Cell by adaptive changing the antennas to form the serving Group Cell. So, cell-edge effect is terminated to enhance the performance of cell-edge and guarantee cell-edge user data rate. This is also the basic requirement of 3GPP LTE and IMT-Advanced. For the user, the handover by Slide Handover strategy is not the user handover among antennas, but antenna choosing for always best user experience through the adaptive change of serving Group Cell.

But for Slide Handover, the handover rules or principles of adding new antennas and replacing or releasing existing antennas are based only on the pilot strength of each antenna. This will constrain the performance of Slide Handover, so, the principle and rule need to be proposed.

In [5, 6], Maximum Utility Principle Access Control method is proposed for multiple access point distributed network architecture, which can maximize the usage of limited system resources with guaranteeing access users' QoS requirements. MUPAC use Dijkstra's Shortest Path Algorithm and Maximum Utility Principle to represent the cost and utility in access control process. For Slide Handover, the Maximum Utility Principle can also be deployed with adaptive revisions.

This handover process is Slide Handover, as detailed shown in Figure 7 with highway environment [11], by which the users are always staying in the centre of Group Cell and the cell-edge effect can be eliminated. When the mobile moves at rapid speed, the size of the slide window will become larger so as to keep up with the movement of the MT and decrease the number of handover times. When the speed of mobile is relatively slow, the size of the slide window will become smaller to reduce the waste of resource. If the MT changes its moving direction, the direction of slide window would change at the same time.

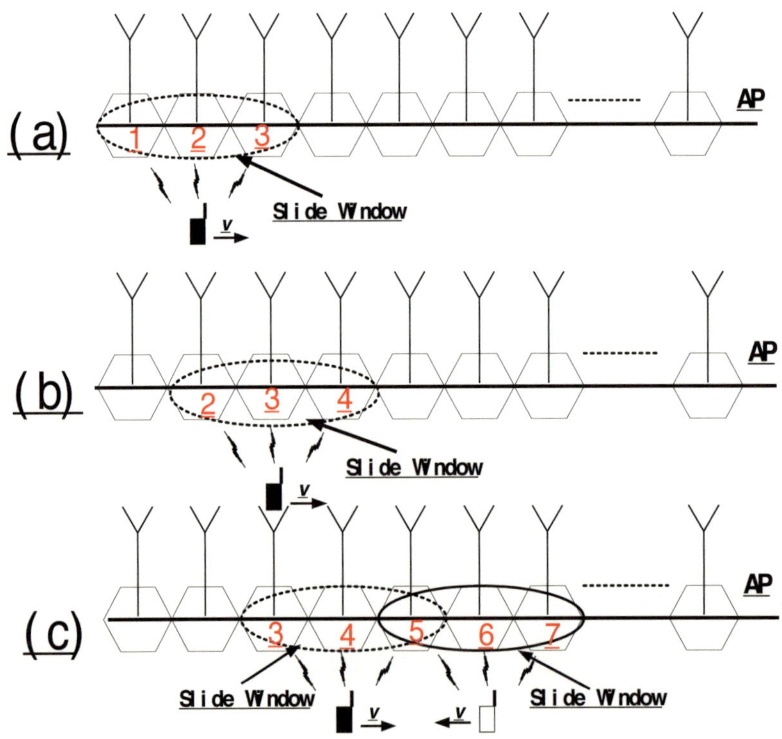

Figure 7. Slide Handover process in highway environment

Slide Handover strategy can complement the handover mainly in physical layer adaptively as AE updated, while the resources will be reserved and signalling in Layer 2 and Layer 3 will not be involved. After the handover, the handover result will be sent to the Layer 3. Slide Handover in Group Cell could be seen as only Layer 1 handover or AE scheduling, where AE is just a kind of radio resource in the multiple access point communication system.

Therefore, Slide Handover can effectively avoid the complicated interchanges of Layer 3 signalling among antennas inside APs and the latency of handover inside APs will be decreased dramatically.

3.2. Maximum utility principle slide handover strategy

Maximum Utility Principle Slide Handover strategy deploys the Maximum Utility Principle into the handover as the rule for adding, replacing and releasing AEs. The principle for adding, replacing and releasing AEs is determined by the utility of each AE to the users. The utility is defined with consideration of both the gain for handover user and the influence to the other users already in the system. Based on Dijkstra's Shortest Path Algorithm, the construction of current serving Group Cell of AEs guarantee the users' QoS requirements with the least numbers of AEs to improve system resource efficiency. Suitable AEs are chosen to form the serving Group Cell by their utility with Maximum Utility Principle. In fact, the Maximum Utility Principle corresponds to the shortest path in Dijkstra's Shortest Path Algorithm. In the process of accessing new user and allocation resources, the optimal choosing multi-AEs to serve one user can be accomplished by adding AEs to user's current serving Group Cell step by step to fulfil the users' QoS requirements. Dijkstra's Shortest Path Algorithm is used in the choosing of AEs by its nature of finding the "shortest path". In this process, the "shortest path" represents the least cost in the access, which is also the maximum utility of adding AEs. This is the key points of Maximum Utility Principle Access Control method.

In Slide Handover strategy, we can still use the Maximum Utility Principle as the rule for adding, replacing and releasing AEs by appropriate definition of Utility Function. The Utility Function in Slide Handover is constructed with considering current system load, resource occupying and so on. The Utility Function has two objectives. One is used to add new AEs to current serving Group Cell with Maximum Utility Principle. The second is to releasing existing AEs in current serving Group Cell because their relatively lower utility. And the third is to select or negotiate resources allocated to the handover users with Maximum Utility Principle.

The utility function has two aspects, including the gain of AEs added in current serving Group Cell and the deterioration for other users in the system. Formula (6) and (7) can still be used in the Slide Handover process.

The detailed implementation steps of Maximum Utility Principle Slide Handover are shown as follows.

1. User initiates handover attempt when its receiving SINR of current Group is below the handover threshold.

2. AP obtains the users' receiving pilot strength of current serving Group and other AEs.

3. Based on the information in step 2), AP calculates the utility of each AE and available resource by utility function and chooses the strongest new AE with the Maximum Utility Principle to add in the serving Group Cell. And the weakest existing AE with the Maximum Utility Principle is replaced and released. This step needs to guarantee the new AE and current serving Group Cell to use the same resources.

4. AP obtains the users' receiving SINR of new serving Group and compares it with the handover threshold. If current serving Group can provide adequate QoS to the user, the handover process accomplishes successfully. Vice verse, the handover users need more AEs added to the current serving Group.

5. AP obtains the users' receiving SINR of AEs excluding current serving Group and chooses new AE with maximum utility to add it into the serving Group or replacing existing AEs. This step also needs to guarantee the new AE and current serving Group Cell to use the same resource. The utility function includes the penalty of resource changing. Then, goes to step 4.

With the Utility Function defined above and Maximum Utility Principle, the Slide Handover can solve the problem of how to choose AEs in the handover process with suitable AE numbers and efficient QoS guarantee. Maximum Utility Principle Slide Handover can maximum the usage of limited system resource and guarantee the handover users QoS. With the advantages of Slide Handover in the physical layer exchanging signalling as AE selection inside AP, Maximum Utility Principle based Slide Handover will get further performance improvement.

3.3. Performance evaluation of slide handover strategy

For the performance evaluation and analyses, traditional Slide Handover strategy with only pilot strength or SINR threshold as the handover rule is taken for performance comparing with Maximum Utility Principle Slide Handover based on Group Cell architecture. Maximum Utility Principle Slide Handover chooses AEs and replaces or releasing AEs according to the Maximum Utility Principle. System-level simulation is adopted to evaluate these two methods by comparing the handover success rate, drop times and throughput in handover. The simulation parameters and setting are shown in Table 2.

Parameters	Setting
Traditional inter-site distance	$500\sqrt{3}$ m
Group Cell inter-antenna distance	500m
Carrier Frequency	5.3GHz
Path gain model	25log10(d)+35.8
Shadow fading deviation	5dB
Total bandwidth	20MHz
Effective bandwidth	17.27MHz
Number of useful sub-carriers	884
Sub-carrier spacing	19.5KHz

Table 2. Simulation parameters and setting

In the simulations, the traditional Slide Handover strategy is denoted as SWHO, the Maximum Utility Principle Slide Handover strategy is denoted as MU-SWHO. The simulation results are shown in Figure 8, Figure 9, Figure 10 and Figure 11.

Figure 8 shows the handover success rate of these two Slide Handover strategies. From the simulation results, Maximum Utility Principle Slide Handover strategy has obvious improvement over traditional strategy because of the rules presented by utility rather than the pilot strength of AEs only. The utility not only contains the gain to the handover users, but also the influence to other users in the system, which makes AEs and resource use more effectively. In some extent, Maximum Utility Principle Slide Handover strategy has the effect of interference mitigation. So, the handover success rate of new strategy is further better.

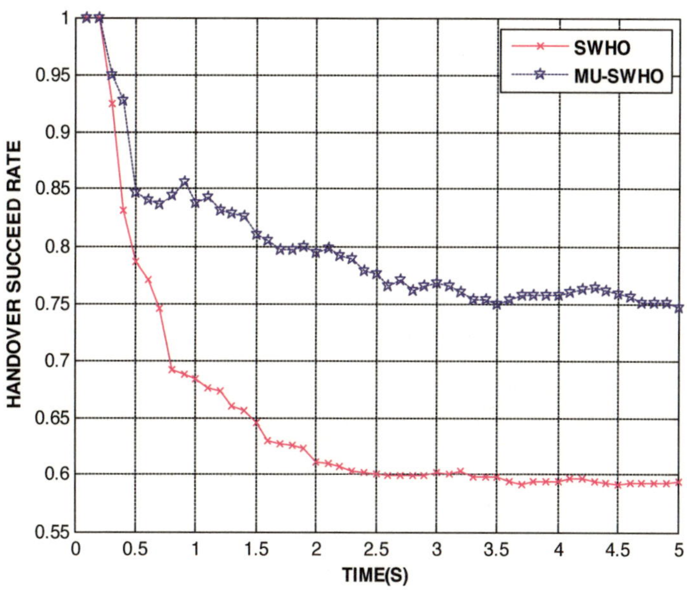

Figure 8. Handover success rate

Figure 9 shows the system throughput of these two Slide Handover strategies. Maximum Utility Principle Slide Handover strategy also has improvement at system throughput in handover process. The reason is that the users in handover choose the best AEs by Maximum Utility Principle, which improves system throughput and also mitigates the interference. The system can achieve optimization.

Figure 10 shows the drop rates of these two Slide Handover strategies. Maximum Utility Principle Slide Handover strategy still has improvement over traditional strategy. Because Maximum Utility Principle Slide Handover strategy chooses AEs based on the users' QoS guarantees and always tries to use as little as AEs to accomplish handover. This will avoid users drop in handover process and improve the resource efficiency with more resources saved for the other users.

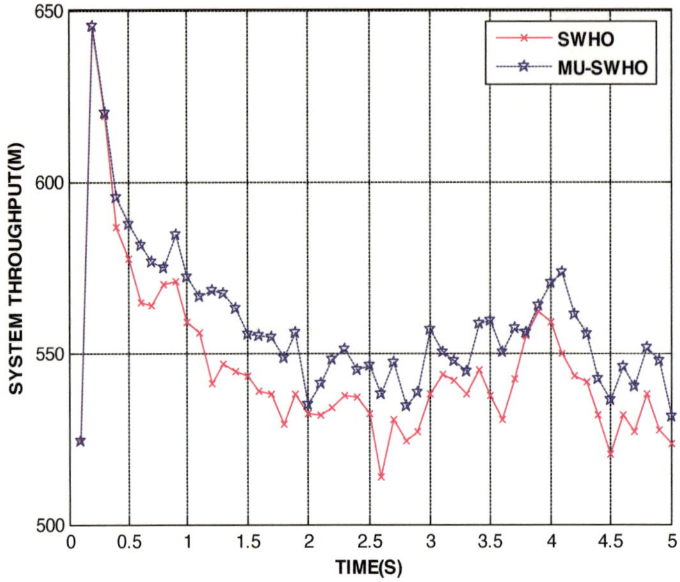

Figure 9. System throughput

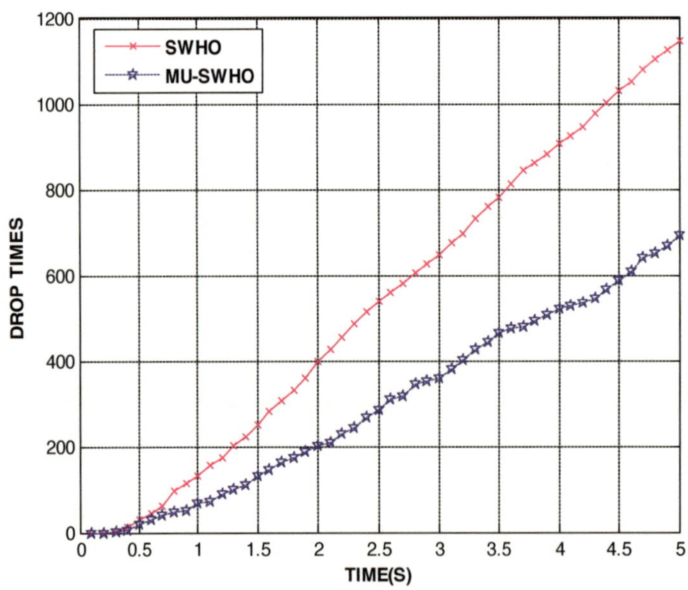

Figure 10. Drop times in handover process

Figure 11 shows the average SINR of AEs inside handover serving Group Cell by these two Slide Handover strategies. By Maximum Utility Principle Slide Handover strategy, the

http://dx.doi.org/10.5772/55401

antennas in the serving Group has better utility than traditional strategy because Maximum Utility Principle only chooses the antennas which have the maximum utility to add in the Group. This will further improve the resource efficiency with more resources saved for the other users.

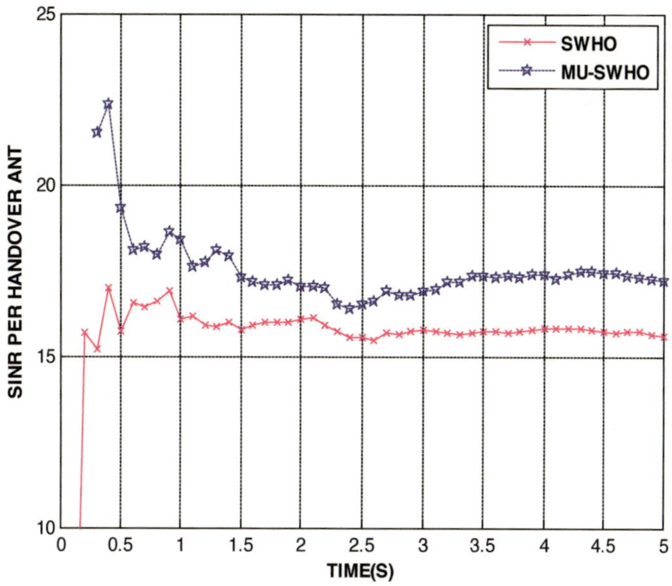

Figure 11. SINR per AE in handover process

4. Summary

Maximum Utility Principle was proposed for multiple access point cooperative communication in vehicle environments based on Dijkstra's Shortest Path Algorithm, which is applied in Access Control method and Slide Handover strategy to improve the performance of mobile communication especially for high speed mobility scenario.

For Maximum Utility Principle Access Control scheme, this chapter presented two improvements for MUPAC with scheduling algorithms. With combination of scheduling and access control strategy, Throughput Targeted-MUPAC and Throughput and Fairness Targeted-MUPAC can get better performance of system throughput and user fairness with appropriate resource utility efficiency to accommodating different situation of system load and access users. Performance evaluation and analyses verify the merits of TT-MUPAC and TFT-MUPAC algorithms in improving system throughput, accessing success rate and user fairness.

For Maximum Utility Principle Slide Handover scheme, this chapter introduced new rules of adding new access points and replacing or releasing existing access points. The Utility Function in the Slide Handover and steps for handover were described in this chapter and system-level performance evaluation was provided to verify the merits of handover successful rate, drop times and so on.

There will still be more challenges for higher mobility vehicle environments, such as the user QoS guarantee and optimal resource allocation and scheduling. More research outcomes need to be achieved.

Acknowledgements

This work is supported by projects of Natural and Science Foundation of China (61001116, 61121001).

Author details

Xiaodong Xu*

Key Lab. of Universal Wireless Communications, Ministry of Education, Beijing University of Posts and Telecommunications, Beijing, China

References

[1] 3Gpp, T. R. Requirements for Evolved UTRA (E-UTRA) and Evolved UTRAN (E-UTRAN), (2005).

[2] 3Gpp, T. R. Requirements for further advancements for Evolved Universal Terrestrial Radio Access (E-UTRA) (LTE-Advanced), (2007).

[3] 3Gpp, T. S. Evolved universal terrestrial radio access (E-UTRA) and evolved universal terrestrial radio access network (E-UTRAN), (2008).

[4] 3GPP REV-080030, Ericsson "LTE-Advanced- LTE evolution towards IMT-Advanced Technology components," 3GPP TSG RAN IMT Advanced Workshop, Shenzhen, China, April (2008)., 7-8.

[5] Xiaodong XuChunli Wu, Xiaofeng Tao, Ying Wang, Ping Zhang, Maximum Utility Principle Access Control for Beyond 3G Mobile System, Wireless Communications and Mobile Computing, Journal of Wiley, Aug. (2007)., 7(8), 951-959.

[6] Xiaodong Xu Dan, Hu, Xiaofeng Tao Access Control Strategy for Multi-antenna Cellular Architecture with Scheduling, ICONS, Nov. (2011).

[7] Xu Xiaodong, Hao Zhijie, Tao Xiaofeng, Ying Wang, Zhongqi Zhang Maximum Utility Principle Slide Handover Strategy for Multi-Antenna Cellular Architecture, IEEE 68th Vehicular Technology Conference (VTC) (2008). Fall, Sept. 2008., 21-24.

[8] Zhang PingTao Xiaofeng, Zhang Jianhua, Wang Ying, Li Lihua, Wang Yong. The Visions from FuTURE Beyond 3G TDD, IEEE Communications Magazine, Jan. (2005). , 43(1), 38-44.

[9] Dijkstra, E. W. A note on two problems in connexion with graphs. Numerische Mathematik, (1959). , 269-271.

[10] Xiongwen Zhao, Kivinen J, Vainikainen P, Skog K. Propagation characteristics for wideband outdoor mobile communications at 5.3 GHz, IEEE Journal on Selected Areas in Communications, (2002). , 20(3), 507-514.

[11] Tao Xiaofeng, Dai Zuojun, Tang Chao Generalized cellular network infrastructure and handover mode-group cell and group handover. ACTA Electronic Sinica, Dec. (2004). , 32(12A), 114-117.

Chapter 6

Distributed Trust and Reputation Mechanisms for Vehicular Ad-Hoc Networks

Marcela Mejia and Ramiro Chaparro-Vargas

Additional information is available at the end of the chapter

http://dx.doi.org/10.5772/55453

1. Introduction

The parallel evolution of automotive industry and anywhere/anytime communication mechanisms has defined a revolutionary moment for intelligent driving concept. Vehicular Ad Hoc Networks, also known as VANETs, exploit the capability of wireless communications protocols to confidently disseminate sensitive information among peers, restricted to a particular geographic area, as well as the advanced safety-assets, on road sensors and sophisticated driving-assistance features of top-generation vehicles. Such a complement are not only oriented to offer a more pleasant in-car experience for driver and passengers, but also are intended to introduce autonomous and efficient ways to prevent hazardous situations on roads. Then, VANETs deployment should follow a primary task based on reliable message handling for sharing traffic conditions, weather variables, driving assistance, navigation support, entertainment content multicasting and even spurious notifications [1], [2].

Regarding some previous approaches over vehicular networks, Vehicle to Infrastructure (V2I) and Vehicle-to-Vehicle (V2V) are two well-know concepts for their implementation [3], [4]. The former claims for a centralized control entity dedicated to information processing and decision making. Although, given the dynamic nature of the involved communications nodes, the ubiquity along the road becomes a mandatory requirement. This might implies a high-affordable effort, due to the demanded deployment of extended communication infrastructure road-sided [5].

On the other hand, the introduction of inter-vehicle communication takes advantage of partially decentralized and self-controlled characteristics from mobile nodes. Such a distributed scenario forms an interim delegation of controlling role, while the privilege to command the network is granted based on the possession of sensitive information to be shared among peers. Correspondingly, V2V communication inherits typical performance hurdles and security threats from distributed mobile networks [6], [7]. Consequently, particular attention should be given to node authenticity, message integrity and reputation features, aiming to promote an integral trust solution [8]. Since an important set of messages contains critical information for the vehicles involved, selfish or malicious behavior should be diminished as possible. On this way, strategies conducted in Mobile Ad Hoc Networks (MANETs) research context are prone to be addressed in VANETs setups as well [9], in order to contrast results from real and simulated scenarios regarding native vehicular network protocols and ad hoc networks' tailored approaches.

The present chapter is dedicated to the review of trust and reputation models for VANETs from different technical, technological and performance perspectives. Besides, novel approaches are introduced, given their capabilities to tackle some of drawbacks and downsides of current approaches [10].

The achievement of trustworthiness among peers infers some inherent challenges owing to vehicles' high speeds and ephemeral associations. For this reason, an initial key issue is directed to guarantee the verification and processing of incoming information in a real-time basis. Likewise, the large population of automobiles in urban streets or suburbs highways at specific time periods affects network channel saturation. Then, some efforts should be oriented to turn the implicit protocols into scalable mechanisms and selective message forwarding. Moreover, VANETs' dynamic properties lead to assume the necessity of decentralized and self-controlled infrastructures appealing to unique and short-term acquaintances among peers. However, hybrid or combined models are subject of discussion in the upcoming sections as feasible VANETs environments. Additionally, VANETs incidents can be distinguished as informational, warning and critical events; in any case spatiotemporal descriptors (i.e. location and time tags) must be properly processed according to priority, life span and certainty aspects. Finally, the adoption of mobility models is directly related with the efficiency of the trust and the reputation system, whereas the accuracy degree of transit patterns influences simulations' results and further implementations in real scenarios [11].

The upcoming sections discuss some remarks on trust and reputation management systems, as well as, trustworthiness models based on diverse mechanisms for scalability, privacy, entity management, content reputation, forwarding, rewarding and so on. Afterwards, innovative models are described in order to establish their original contributions for challenges overthrowing and general VANETs development.

2. Trust and reputation for VANETs in scope

Pursuing multi-featured trust and reputation models, issues of scalability, security, performance and sustainability should be addressed in VANETs. Inasmuch as multi-featured trust and reputation models are pursued, topics from generalization, security, performance and sustainability are addressed. The following criteria enclose the requirements and guidelines in order to accomplish outstanding trust and reputation models [12], [13].

2.1. Low complexity

The interactions among VANETs' peers are characterized by a circumstantial occurrence at high speeds and short time periods (ephemeral acquaintances). During this timeframe, the sender automobile should transfer reliable information within its influence radius. The

http://dx.doi.org/10.5772/55453

computation of trust and reputation metrics should be executed by the master algorithm under low complexity constraints, i.e. cost-efficient processing, fast-access memory, improved transmission rates and effective throughput.

Hardware and software specifications related to processing and storage are conceived not only to fulfill single-threaded duty cycles at high frequencies, but also to support eventual burst of messages from different sources at given time instants. Similarly, data transmission rates need to rely on recognized standards and specifications in charge of modulation schemes, media access and packet routing. Even so, the overhead linked to the throughput portion should be kept at minimum, since excessive encode/decode of protocols headers leads to critical delays for the reception of meaningful driving information.

2.2. Scalability

Scalability should be understood as a fundamental condition for a trust and reputation mechanism in any application. Moreover, the occurrence of incidents or events on the road could be triggered by multiple mobile nodes, as well as single or minor set of vehicles. Either way, the system should be prepared to conveniently process any incoming information at bursting rates, preventing increased packet drop or perceptible latency at decision time.

An initial approach implies the improvement of system's physical capabilities to handle plenty of concurrent messages, and thus, reducing the probability of missing messages or extended processing cycles for warnings and alerts posting. In general, the fulfillment of requirements by the overestimation of resources seems to be the most immediate, but rarely, the most efficient solution. Therefore, alternative methods should be employed to chase model's top-performance, while low complexity and simplicity are preserved. For instance, solutions based on reputation records could allow selective forwarding and reception of incoming and outcoming information, respectively. So that resources are allocated to nodes that have displayed a good behavior within the network. Of course, prerequisites must be taken into account to promote the success of this model; such as precedent behavior awareness and reputation history access. Balance-oriented solutions claim for fair resources disposal, accompanied by the definition of conservative thresholds around incoming packets upper limits, relay nodes allowance and messages intervals designation.

2.3. Sparsity

Some VANETs' environments can be described as dense populations of vehicles confined to relatively tight geographic areas, e.g. peak hours in urban zones. Other environments, like interstate or international highways are characterized by extensive trails with minimal density of automobiles. This situation makes difficult the application of multihop routing protocols from sources of events to spots of interest. Consequently, trust and reputation models should not be based on the evaluation of peer interactions, such as lists of relay nodes and reputation scorecards. Instead of this, these models should take into account the scarcity of information in the VANET. So, decision making algorithms and thresholds settings need to be as flexible as possible to process these messages. Even more, they should provide valuable data to the onboard driver and eventual adjacent mobile nodes, instead of attempting an unbiased execution of the trust and reputation mechanism.

2.4. Security and privacy-related

Undoubtedly, the extension of a security middleware for VANETs is understood as a major concern, relaying either on centralized or distributed schemes. Generally, every different approach aims to shield the communication infrastructure against potential security threats and vulnerabilities [14]. For example, one usual threat is represented by malicious mobile nodes, which might attempt to disseminate false or corrupted information by making use of resources and the communication channel, perversely. Dissimilarly to other network contexts, misbehavior of a single node might imply severe or even lethal consequences to benign peers; whereas the injection of deceiving content is prone to conclude in inattention of vital recommendations or commitment of undesired actions. The author Zhang in [13] describes some common attacks that represent a critical challenge for trust and reputation mechanisms. Some of them are briefly introduced as follows.

- Newcomer Attack: Specially applied by mobile nodes with an undesirable cooperative behavior. By the registration of a new identity in the trust and reputation system, the malicious node attempts to delete its negative history to gain the attention of adjacent nodes as a freshman.
- Sybil Attack: Consists of creating multiple and fake identities (pseudonyms) by a single malicious entity. Thereafter, a pass-through is granted to the false information by pretending peers, who attempt to detour network resources to their own benefit or collapse general message distribution.
- Betrayal Attack: Appealing to a hypocrite strategy, a malicious peer formidably cooperates within the network until high reputation and trust scores are received. Suddenly, its behavior turns inadequate by propagating deceiving content, while it takes advantage of the influence over the mobile neighbors.
- Inconsistency Attack: Related to the previous attack, a vehicle oscillates between benign and malicious behavior at different periods of time. The intention of this attack seeks to destabilize the trust and reputation mechanism by entering chaotic records to the observed interaction among mobile nodes.
- Collusion Attack: Also known as conspiracy attack, a group of malicious peers subscribes a dishonest coalition to generate false information to the network from multiple points. Keeping relatively outnumbered agents, the attack may also affect or even collapse the content distribution system among vehicles.
- Bad-mouthing/Ballot Stuffing Attack: Following the line of conspiracy attacks, a set of malicious nodes gains access to the network with a cooperative behavior. Once, rating or feedbacks about other adjacent peers are requested, inaccurate opinions are provided. Attempting to unfairly increase reputation of suspicious nodes (ballot stuffing) or unfairly decrease reputation of benign entities (bad-mouthing).

Defense mechanisms for the aforementioned attacks are regarded from the universal framework of information security for general data networks [15], including AAA protocols, symmetric, asymmetric systems, cryptographic key management, etc. However, absolute solutions have not still been met; e.g. the absence of key distribution mechanisms may lead to the interception of shared secrets by unauthorized entities in symmetric key deployments. On the other hand, the asymmetric cryptosystems may compromise the public keys'

http://dx.doi.org/10.5772/55453

distribution by sophisticated techniques of replacement or theft of network identities and traffic information. Therefore, keys authenticity assurance turns into a vital security matter. So, the deployment of credentials, from now on called certificates, allow us to bind the public key to the owner's name and a trusted third party, designated as Certificate Authority (CA). Of course, the generation, management and revocation of certificates may become a complex endeavor when the number of network entities tends to increase. For this reason, a Public Key Infrastructure (PKI) rises as a solution to handle major aspects, such as certificates' lifecycles, issuance, distribution, suspension and revocation. Although, further considerations should be applied to adapt PKI to a partial or total distributed topology in VANETs environment.

2.5. Independence of mobility patterns

Mobility pattern is a central point of the discussion about VANETs topologies. Even more, mobility pattern is a crucial issue in ad hoc networks research, given its direct relation with protocols, models and performance analysis. The definition of restrains for the transit, environment variables and interaction rules among entity nodes, determines a set of standard rules for a correct simulation and performance analysis of VANETs. In VANETs context, a group of mobility patterns has been proposed intending to assemble an universal playbook to guide the simulation of different vehicular models, including trust and reputation. Those patterns aim to set parameters related to automobiles density, traffic area, traffic lights and stop signs existance, average speeds, block and streets disposition, weather conditions, overtaking chances and any other variable that can emulate a real vehicular scenario [11].

Despite of being so attractive, the idea behind of a trust and reputation model entirely independent of mobility patterns, there are reasonable and heuristic assumptions that might lead to disregard this possibility. Instead of designing a unique trust and reputation model for any possible mobility pattern, major efforts should be conducted to adapt variations of the mechanism to consider some of the most recurrent patterns in urban or rural areas. Consequently, models' structure should include degrees of freedom that easily permit alternative resources allocation and algorithmic shortcomings, depending on detected transit, environmental and interaction parameters.

2.6. Trust and reputation decentralization

The distributed and self-controlled feature of VANETs is an extensively accepted concept, given the mobile-oriented dynamic of the nodes. Thus, lots of research works refer to message relay, security assurance, privacy adoption and trust/reputation establishment toward decentralized deployments. The interaction among peers is the base of trust and reputation construction [16]. Random peer-to-peer acquaintances allow the establishment of trustworthiness relationships by references at first hand, i.e. the definition of reliable nodes depends on direct observations by the sensor-equipped vehicle on road. It is expected to assign a greater confidence valuation to those nodes, once the event or incident is corroborated by an observer peer. Complementary, referral mechanisms are introduced to trust and reputation systems in order to optimize the convergence time and awareness status. That means, an interested node might request recommendations or opinions from adjacent nodes to assign a trust value to peers out of the scope. Also, it is expected to set a lower confidence degree to unreachable vehicles, ought to the lack of personal certainty about its reliability. Achieving a complete decentralization is not easy due to slow convergence, delays in peer status update, network changes, etc. Indeed, the uncertainty about short-term or long-term encounters among vehicles hinders the assembling of trust information over the whole network. Because of this, there are proposals to introduce some controversial entities in VANETs, the Road Side Units (RSU). They would coordinate the gathering of information at infrastructure based concentration points, in a centralized approach. Due to a hard-wired interconnection (e.g. optic fiber rings), RSUs are capable to share information at significant transfer rates to be transmitted immediately to vehicles within their coverage. In spite of breaking apart the decentralization concept, coexistence of both mechanisms for trust and reputation establishment are still widely studied [17], [18].

2.7. Confidence measure

Even though, trust and reputation mechanisms have been able to carry on trustworthiness valuations successfully. Some additional metrics need to be applied, in order to introduce "Quality Assurance" (QA) into VANETs' models. The knowledge level about mobile peers may not be always associated to highly consistence databases, due to several facts. For example, minimum requirements for algorithm computation, lack of memory, unavailability of data or simply owners' and manufacturers' discretion.

For this reason, confidence measure stands out as a statistical QA parameter to assess the accuracy degree of modeled trustworthiness values [13]. A scoped vehicle makes use of this measure to decide how useful the incoming content could be, according to particular circumstances on the road. Such that, an automobile with low confidence measures about neighbors and events is encouraged to seek alternative referees to improve its levels of certainty; but the same vehicle who faces scarcity of information sources has no better option than supports its decision on the current confidence metric. Then, exactly the same parameter can be easily ignored at one particular situation and overvalued throughout in another.

2.8. Event description and spatiotemporal specification

There are various types of events to happen on roads, highways or even quiet urban lanes. The description level plays a fundamental role within trust and reputation models to assign weights and priority tags to incoming information, according to messages' description. The distinction among informational, warning and critical events needs to be clearly specified and managed by all peers. Not only to suggest a suitable action in consequence with the input data, but also to incentivize with proportional rewards to the forwarding nodes. It means, nodes that are willing to cooperate wisely with life-saving messages over meaningless broadcasting.

Likewise, location and time parameters exhibit further benefits in trustworthiness assessment than merely timestamps accounting features. High-dynamic essence of VANETs infers real-time spatiotemporal feedback about vehicles walk-through, expecting minor location and launching efforts of authorities, when incidents occur. Once a major event is reported, mobile nodes that are closer located and sooner reckoning might offer a greater accuracy of transmitted information, as well as, the kind of assistance required. In this case, advantageous geolocation of reporter nodes should be also treated with higher trust values and subsequently generous reputation scores should be awarded, if information is confirmed.

Summarizing, Table 1 depicts the criteria for trust and reputation models in VANETs with their corresponding current development status, according to the related research work.

Feature	Pending	In progress	Covered
Low Complexity		×	
Scalability			×
Sparsity	×		
Security and privacy-related		×	
Mobility patterns independent			×
Trust and reputation decentralization			×
Confidence measure	×		
Event and spatiotemporal specification			×

Table 1. Criteria for trust and reputation models in VANETs.

3. Discussion on trust and reputation models

At this point, a rich collection of challenges about trust and reputation modeling for VANETs has been introduced. Now, a review of existing mechanisms is prepared, attending to challenges, design criteria, implementation hurdles and support insights. Depending on particular VANET context, several models are explored, appealing to heterogeneous techniques over infrastructure, interaction and trustworthiness assessment matters. In what follows, we will review some of them.

3.1. Content Reputation System - CoRS

CoRS protocol is a content-based message reputation model, i.e. message content is the primary analysis object for trustworthiness computation [17]. Nevertheless, additional features are applied for node authenticity and information integrity based on cryptographic mechanisms. PKI infrastructure and threshold cryptography are the components employed by CoRS to assess reliability of mobile nodes, while the protocol focuses on content reputation, exclusively. The former demands a valid pair of cryptographic keys for each participant in the VANET, which are usually provided by the CA. The distribution of digital certificates includes a public and a private key, which are meant to establish a secure channel among peers for the exchange of messages. As long as ciphersuites are in charge of encoding and decoding tasks at each side. In the same manner, digital certificates are engaged to perform authentication procedures in order to validate peers' identity, as a first step.

On the other hand, the threshold cryptography system performs a critical role within the content reputation model. Despite of being an old-fashioned concept in information security literature, its validity remains quite bearable. The basic concept behind cryptographic threshold states a mechanism to divide into n chunks a whole data packet P, and then, a previous knowledge of c or more chunks will permit the complete packet P reconstruction [19]. But only, defining up to $c-1$ chunks will lead to a certain data undisclosed; so polynomial interpolation is applied to compute the initial threshold pair (c, n). From that, the theorem claims that given c different points on the two-dimensional plane, such that $(x_i, y_i)\ \forall\ i \in [1, c]$, one and only one polynomial $f(x) | f(x_i) = y_i$ with degree $c-1$ exists. The remaining chunks of the data packet P can be found with Equation 1

$$f(x) = a_0 + a_1 x + a_2 x^2 + a_3 x^3 + \cdots + a_{n-1} x^{c-1} \quad (1)$$

where a_0 represents the whole data packet P and the other coefficients a_i are chosen randomly. To summarize, the definition of c chunks and their corresponding indices conducts to find the coefficients of $f(x)$ and recover data packet $P = f(0)$ by interpolation computation.

The threshold cryptography is applied in threshold signatures generation, which is the core machinery of trustworthiness assembly in CoRS. This cryptosystem introduces digital signatures for distributed environments by the usage of three major components: 1) a threshold public key K_{pub}, 2) a certificate C and 3) a key share K_{share}, which is a partial private key. Such that, only the combination of cooperative nodes with their respective K_{share} allows digital signature verification, and thus, content validation. The management of K_{share} is delegated on a CA or share dealer; which should cope with the determination of the minimum number of K_{share} required to successfully sign a message, as well as, at least one share delivery to each mobile node. Figure 1 shows players and workflow of threshold signature for distributed reputation systems.

Once, the share dealer has generated and distributed K_{share} among registered vehicles, a random combiner executes an algorithm to integrate partially signed data chunks; whose outcome produces a valid digital signature to sign the message. So, the resulting signature can be verified using the corresponding certificate, whilst individual shares are kept in secret during combination process. Figure 1 depicts the threshold signature procedure for CoRS protocol.

Having all the pieces into place, we will proceed to explain the involved phases in CoRS mechanism.

3.1.1. CoRS initialization

The CA or share dealer generates the required certificates (including public cryptographic key K_{pub}), and shares K_{share} for n mobile nodes. Attempting to avoid a constant reset of the system every time a new vehicle registers into a particular VANET context, the relation between the number of nodes and shares is kept $N_{nodes} \gg N_{shares}$. That implies, more than one automobile is using the same K_{share} to partially sign the message. Gathering those components, each mobile node is prepared to perform authentication, data protection and message integrity, before content reputation system is taking place.

3.1.2. CoRS implementation

The protocol starts to detect an event or incident at some mobile node, known as *generator* (stage 1). Before it sends an information message msg, some support data is requested from the nodes located in the *reputation area*; even though the adjacent nodes very likely detect the same event, they are denoted as *verifiers* (stage 2). Then, *generator* produces a reputation request, which is signed using its personal private key K_{priv} and it is sent to the *verifiers* with the message itself. At *verifier's* side, the reputation request is validated employing the cryptographic elements. Further, content is verified, trying to check the several pieces of

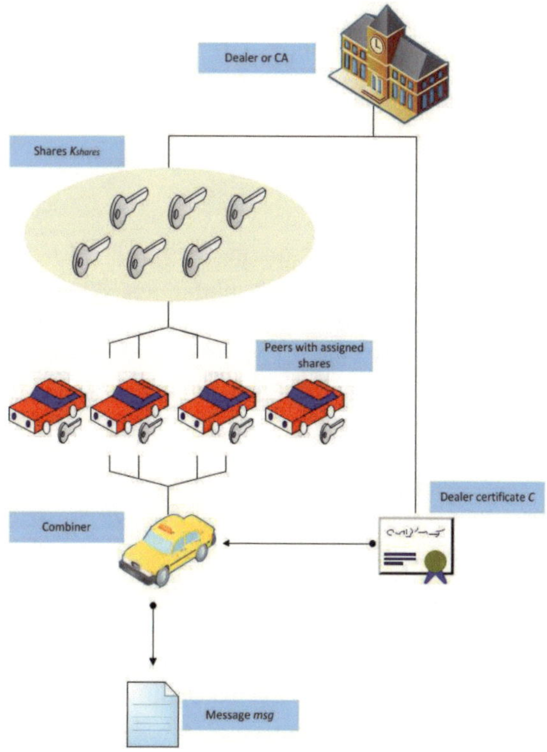

Figure 1. Threshold security for CoRS protocol.

information to evaluate the values for the reputation metrics r_m (stage 5). Such metrics serve to rate candidate events and message settings, depending on the most suitable category according to the following information.

- Event: In the case that the *verifier* has detected the same event as the *generator*, the former has the right to examine the message *msg* later. If not, the protocol workflow is terminated and the reputation request is not signed with its own K_{share}.
- Event detection time: The *generator* sets the message's timestamp as same as the event detection time. When the *verifier* receives a reputation request, its timestamp should not be larger than a predefined threshold, so only recent events are processed.
- Location of the event: The *verifier* determines the location of the event included in the reputation request, in order to define if it is identical to the incident observed by itself in its coverage area.
- Location of the node: To check the location of the *generator* in comparison with the potential *verifiers* and the event itself, a *collator* is required to resolve whether the distance is reasonably close.
- Sending time of the request: Eventually, a *collator* is not available at the sending time of the reputation request. Then, a second request will be transmitted with an implicit delay. The difference between event detection and sending time must handle some tolerance to allow further protocol computation.

Once, the reputation r_m information have been gathered to rate the trustworthiness level, a reputation computation is performed as follows.

$$Rep(msg) = \sum_{m \in \Re} r_m \quad (2)$$

For simplicity matters, unit-value scores are granted, i.e. +1 is given to affirmative validations, 0 for neutral and -1 for unsuccessful validations. While, $Rep(msg) \geq 0$ condition is fulfilled, the message *msg* is regarded as valid with a positive reputation and the *verifier* is set free to send a reply to the *generator*. Otherwise, the request reputation is neglected, turning the mobile node into a *denier*. The transmission of the reply message follows the same threshold cryptography guidelines, assuring authenticity and integrity by digital signing with K_{share} (stage 6). Furthermore, the *generator* collects all the incoming messages in order to validate digital signatures, and finds the information associated with the generated reputation request (stage 7). If all parameters of threshold signatures are hold (refer to section 3.1 introduction) (stage 8), the information message *msg* is distributed to other peers outside of the *reputation area* to let them know about the event or incident that is happening on the road.

As optional protocol's add-ons, broadcast manager (stage 3) and DoS protection (stage 4) help out to prevent broadcast storms and irregular content distribution, respectively. The former avoids reprocessing of already sent messages and the latter continuously check a list of banned peers, given its dishonest previous behavior.

The Figure 2 summarizes the stages involves in CoRS protocol workflow, based on player and general actions.

3.1.3. CoRS remarks

Despite of the reliance of CoRS protocol on central entities, the participation of central authorities or dealers represents a well-conducted and sustained strategy for node authentication, message integrity and data confidentiality. However, additional countermeasures are taken to avoid reiterative CA advisory, like threshold digital signature in a distributed mode. Though, the performance of different ciphersuites is not particularly discussed in CoRS framework; the protocol's workflow permits to infer a low complexity nature. By observing the protocol's pseudocode and involved stages, from event detection until information dissemination. CoRS preserves a balance between demand and consumption of resources [17].

Besides of this, threshold cryptography allows to improve scalability and sparsity, due to the bulky generation of shares among peers.

Nevertheless, it is possible for shares to collide during the normal execution of CoRS. Since, the number of nodes is always far greater than the number of shares, the possibility of assigning the same share to more than one automobile is quite plausible. Then, statistical modules may be needed to foresee encounter among peers with the same cryptographic signing material. Such a situation increases the complexity and time response of the general protocol. Owing to the distributed threshold cryptography, a compromising attack from

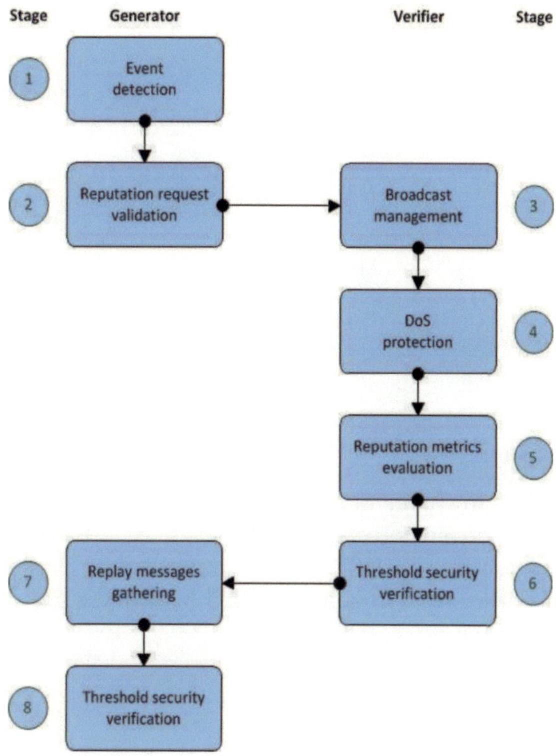

Figure 2. Content Reputation System (CoRS) workflow and stages [17].

one isolated node seems to be unfeasible. But, perverse coalitions among malicious peers (al least equivalent to the number of signing shares) will lead to successful collusion and bad-mouthing/ballot stuffing attacks. Finally, the existence of a role player, known as collator for location and time judging before ambiguities presence, it is more than necessary for the proper protocol walkthrough. Apparently, non-contingent policy is introduced to confidently step over or replace this sensitive requirement, so a high-faulty point is discovered.

3.2. Trust and Reputation Infrastructure-based Proposal - TRIP

TRIP protocol can be labeled as a hybrid trust and reputation model, whereas trustworthiness assessment is entity node- and content-oriented [20]. The decision making starts with a reputation score computation referred by three different possible sources: 1) directly acquainted mobile nodes, 2) referee nodes with indirect contact and 3) centralized authorities, like CA or RSU. Secondly, each mobile node is classified into three different trust levels, represented by fuzzy sets [12]. Considering, nodes' categorization further actions are committed. For instance, absolute data rejection (Not trust), data acceptance but not forwarding (+/- Trust) and message acceptance and forwarding (Trust). Finally, the information message is associated to a particular severity, priority or hazard. Only peers located in the highest trust level are allowed to transmit messages with the most critical severity. Similarly, peers settled in unfavorable trust levels will not find successful acceptance of any kind of message.

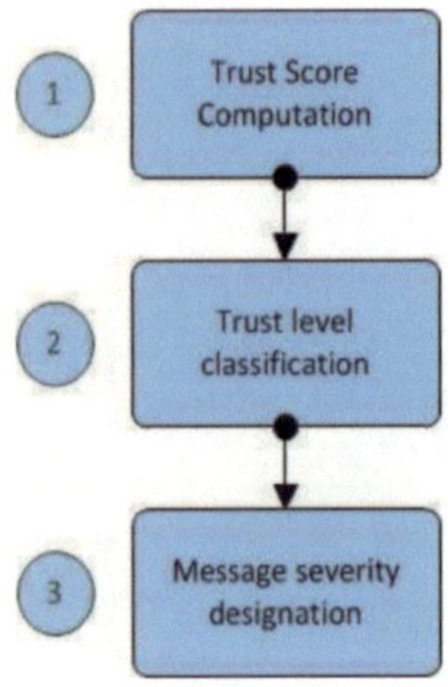

Figure 3. TRIP protocol workflow and stages.

3.2.1. Trust score computation

Let define v_i as the vehicle in charge of assessing a trust score for another vehicle v_j, which sends the message with a current situation on the road. As it was mentioned before, three different sources are empowered by TRIP protocol to issue a reputation score. Equations 3-5 denote the notation given to the calculation emitted by a direct peer, a referee node and a centralized entity, respectively.

$$\alpha_i * Rep_{ij}(t-1) \quad \forall \quad Rep_{ij} \in [0,1] \quad \alpha_i \in [0,1] \tag{3}$$

$$\beta_i \sum_{k=1}^{N} \omega_k * Rec_{kj} \quad \forall \quad Rec_{kj} \in [0,1] \quad \beta_i, \omega_k \in [0,1] \tag{4}$$

$$\gamma_i * Rec_{RSUj} \ \forall \quad Rec_{RSUj} \in [0,1] \ \gamma_i \in [0,1] \tag{5}$$

From Equation 3, a mobile node i has given a reputation score *Rep* about node j at a previous time instant $(t-1)$; and a tunable weight α_i is granted. Accordingly in Equation 4, a set of N peers indexed by k issues a recommendation *Rec* about node j, where ω_k represents the reliability of such referrals and β_i assigns a corresponding weight to the source. Finally, Equation 5 infers the recommendation value generated by RSU to a node j, accompanied by its weight γ_i. All the scores and weights are constrained to be allocated in the interval $[0,1]$. Merging the prior expressions, it is obtained the trustworthiness assessment for TRIP protocol, as shown in Equation 6.

$$Rep_{ij}(t) = \alpha_i * Rep_{ij}(t-1) + \beta_i \sum_{k=1}^{N} \omega_k * Rec_{kj} + \gamma_i * Rec_{RSUj} \quad \forall \ Rep_{ij} \in [0,1] \tag{6}$$

As soon as trust score is calculated, encouragements and punishments are scattered across the reputation area by the confirmation or denial of the reported event. The weights $\alpha_i, \beta_i, \gamma_i$ and reliability index ω_k are subject of increments and decrements, based on the accuracy level in the information disseminated by i^{th} and k^{th} peers at time instants t. Furthermore, the identification of customary malicious nodes can be referred to central authorities or RSUs to compose a black list, which it is employed to ban spurious peers' participation with no further computations.

3.2.2. *Trust level classification*

The analytical value attained by Equation 6 leads to make a decision about the received information. According to the score, the message may be rejected, accepted but not forwarded, or accepted and forwarded. The authors in [12] proposed the fuzzy sets rendered in Figure 4 to associate trust levels and scores.

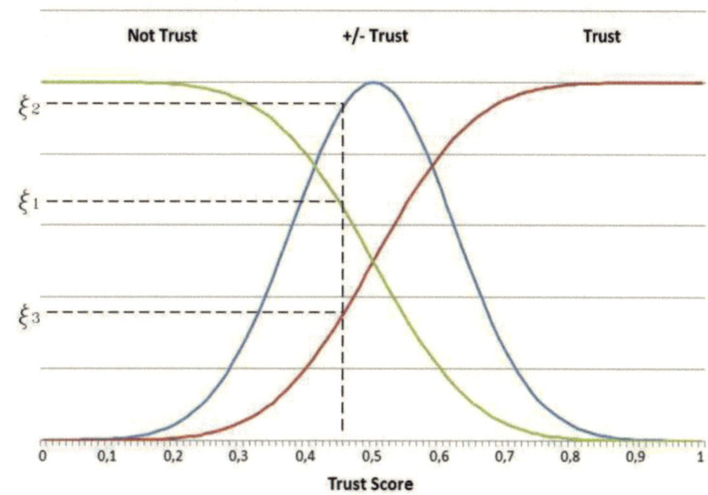

Figure 4. Fuzzy sets for trust levels [12].

To determine which trust level the automobile belongs to, each fuzzy set owns a membership function $\xi_F : F \rightarrow [0,1]$ (see Figure 4). Such that, the probability that a mobile node is placed in trust level TL_k is obtained by Equation 7.

$$P(TL_k) = \frac{\xi_k}{\xi_1 + \xi_2 + \xi_3} \qquad \forall\ \xi_k = \xi_{TL_k}(Rep_{ij}(t)) \tag{7}$$

Given that a spotted vehicle is allocated in *Not Trust* region, its message is immediately neglected and one entry is entered in the black list managed by RSU. On the contrary, if the spotted vehicle is awarded in *Trust* set, its information is processed and shared with the nodes within the coverage area. The shadowing zone, labeled as +/- *Trust* supports message acceptance with no forwarding chance. However, the message acceptance is conditioned to an adaptable probability, designated as follows.

$$P_{+/-T} = \mu_{+/-T} - \mu_{NT} - \sigma_{NT} \tag{8}$$

where $\mu_{+/-T}$, μ_{NT} and σ_{NT} are the +/- *Trust* mean, *Not Trust* mean and *Not Trust* standard deviation, correspondingly. So, the higher $\mu_{+/-}$ becomes and the lower μ_{NT}, σ_{NT} tends, the higher the probability of acceptance $P_{+/-T}$ will be set.

3.2.3. Message severity designation

At the end, a certain severity level is given to each received message; which can be distinguished for important and hazard messages, information warnings and advertisement or less critical content. Therefore, Table 2 sums up the implicit relation between message classification and trust levels assigned to particular nodes.

Severity/Issued by nodes	Trust	+/- Trust	Not Trust
Hazard messages	×		
Information warnings	×	×	
Advertisement content	×	×	

Table 2. Acceptance of messages' severity with respect to trust levels.

3.2.4. TRIP remarks

TRIP protocol makes use of simple but well-assembled instruments in trustworthiness pursuing, by assessing trust scores with weighted polling strategies. Also, fair rewards and penalties are imposed to multilateral information sources, whilst the option of gaining higher rates is present, as long as peers interaction keeps in progress. Moreover, allocation of nodes and messages to different classification scales is based on tunable probabilistic functions, which might be adapted according to VANET context and observed nodes' behavior.

Unfortunately, no effort is intended to ensure the authenticity of the node, peers partial identification or pseudonyms avoidance. Even so, parallel research works in MANETs are proposed by the authors to cope with that. Therefore, the back door for newcomer and Sybil attacks is open, since the protocol by itself does not implement resilient mechanisms against them. In spite of integrating central entities or RSU as collaborative members for the query of malicious nodes' black list, the protocol manages to make of it a dispensable feature.

3.3. Data-Centric Trust Establishment framework - DCTE

For DCTE framework, trustworthiness establishment is sought through content analysis rather than entity nodes' identification. In paper [21], the authors claim: *"data trustworthiness should be attributed primarily to data per se, rather than being merely a reflection of the trust attributed to data-reporting entities"*. The derivation of trust and reputation metrics is focused on the amount of information that can be extracted from reported events. Likewise, multiple sources of evidence are taken into account to assign specific weights, in regard of inherent variables, such as geolocation or time occurrence. Consequently, original data and weighted metrics perform as input variables for a *decision logic* algorithm, which resolves a trust level output.

3.3.1. DCTE definitions

Before going through the details of event-reports evaluation or decision logic techniques, we will introduce some basic definitions for VANET environment contextualization.

- A set of mutually exclusive basic events, denoted as $\Omega = \{\alpha_1, \alpha_2, \alpha_3, \cdots, \alpha_I\}$ can be understood as traffic jam, slippery road, detour section, etc. Similarly, composite events $\Gamma = \{\gamma_1, \gamma_2, \gamma_3, \cdots, \gamma_I\}$ are the unions or intersections of basic events.
- A set of nodes or vehicles, expressed as $V = \{v_1, v_2, v_3, \cdots, v_K\}$ are classified following a system-specific set of node types, $\Theta = \{\theta_1, \theta_2, \theta_3, \cdots, \theta_N\}$. For consistency, let define a function $\tau : V \rightarrow \Theta$ to assign and return the type of a scoped vehicle v_k.
- Let set a *default trustworthiness* for a node v_k of type θ_n as a real value, depending on particular node attributes (e.g. onboard sensor equipment or on road authority member). Such that, every node type owns a unique and consecutive trustworthiness ranking, explained by $0 < t_{\theta_1} < t_{\theta_2} < t_{\theta_3} < \cdots < t_{\theta_N} < 1$.
- Let $\Lambda = \{\lambda_1, \lambda_2, \lambda_3, \cdots, \lambda_J\}$ be the set of tasks related to the protocol or system. Thus, two vehicles (v_1, v_2) with returned types $(\tau(v_1) = \theta_1, \tau(v_2) = \theta_2)$ and known default trustworthiness rankings $t_{\theta_1} < t_{\theta_2}$ are assumed, though it is still possible that v_1 is regarded as more reliable than v_2 in function of task $\lambda_j \in \Lambda$.
- Two input arguments, node type $\tau(v_k)$ and system task λ_j conform *event-specific trustworthiness* function, distinguished as $f : \Theta \times \Lambda \rightarrow [0,1]$, which is invoked to differentiate among nodes of the same type when particular actions are required.
- In terms of security, let introduce *security status* function $s : V \rightarrow [0,1]$, where $s(v_k) = 0$ means the revocation of the node v_k and $s(v_k) = 1$ infers the node legitimacy. Any intermediate value within the interval may be used to characterize scaled security levels.
- At last but not least, let set a *dynamic trust metric* function, expressed by $\mu_l : V \times \Lambda \rightarrow [0,1]$. The index l points out dynamically changing attributes of nodes. Therefore, for every attribute, a corresponding metric μ_l is applied.

3.3.2. DCTE trustworthiness function

The computation of trustworthiness is data-centric or report-oriented. Also, the generated value from the jth report e_k^j is provided by K distinct mobile nodes v_k, which supports on scattered evidence of event α_j. The integration of default trustworthiness, security status, node type and event-specific trustworthiness functions shapes a general trust function, denoted by Equation 9

$$F(e_k^j) = G(s(v_k), f(\tau(v_k), \lambda_j), \mu_l(v_k, \lambda_j)) \quad \forall \; F(e_k^j) \rightarrow [0,1] \tag{9}$$

The obtained weights or trust levels within the interval $[0,1]$ are assessed by a vehicle v_k with respect to every incoming event report from surrounding nodes. Since, the combination of multiple pieces of evidence is one major concern in DCTE, one unique weight is not a confident enough outcome. Henceforth, the composite of various weights related to the same event will conduct to a more robust and reliable decision material. Then, the reports accompanied by their matched weights are transferred to a *decision logic* module that manages to find a definitive action to be taken, like message disposal, conditioned forwarding or full-compliance.

3.3.3. DCTE decision logic

The decision logic module is strongly related to multisensor data fusion techniques. Hence, the performed algorithms are rule-based systems, matching simple polling [22], weighting [23] or statistical procedures. In DCTE context, bayesian inference (BI) and Dempster-Shafer theory (DST) are discussed as candidate data fusion techniques for decision logic implementation.

Bayesian Inference

BI is supported by the well-known Bayes' theorem, where the blended weight of a particular event α_i is defined by the *posteriori* probability of α_i given novel pieces of evidence, $e = \{e_1^j, e_2^j, e_3^j, \cdots, e_K^j\}$ and it is expressed in terms of *apriori* probability $P[\alpha_i]$, as follows.

$$P[\alpha_i|e] = \frac{P[\alpha_i]\prod_{k=1}^{K} P[e_k^j|\alpha_i]}{\sum_{h=1}^{I}(P[\alpha_h]\prod_{k=1}^{K} P[e_k^j|\alpha_h])} \tag{10}$$

From Equation 10, it is assumed that event-reports are statistically independent; i.e. the receiver node is unable to figure out dependencies in reports from different vehicles, which is rational whereas such an information is not provided within reports. Thus, $P[e_k^i|\alpha_i]$ represents the probability that k^{th} report confirms the event α_i, given that α_i occurred. By recalling Equation 9, probability and weights of reports can be equalized as:

$$P[e_k^i|\alpha_i] = F(e_k^i) \tag{11}$$

In case of detecting, a further report that does not confirm the event α_i, given that α_i occurred. It would correspond to a malicious or deceiving node, who is reporting a fake event. So, the probabilistic complement is denoted by Equation 12.

$$P[e_k^j|\alpha_i] = 1 - P[e_k^i|\alpha_i] = 1 - F(e_k^i) \quad \forall\ i \neq j \tag{12}$$

Dempster-Shafer Theory

One characteristic of DST remains in the tractability of evidence, even though there is lack of information upon reported events. By the occurrence of two clashing events, measured uncertainty about one may serve as supporting evidence for other. In comparison with BI, the probability is replaced by an uncertainty interval, upper bounded by *plausibility* and lower bounded by *belief*. The belief value assigned to an event α_i is provided by the Kth report as the sum of all basic belief assignments $m_k(a_q)$, where a_q collects all basic events that integrate the event α_i. Correspondingly, the plausibility value of an event α_i is the sum of all evidence that does not refute such event. Thus, *belief* and *plausibility* are described by Equations 13 and 14, respectively.

$$bel_k(\alpha_i) = \sum_{q:\alpha_q \subset \alpha_i} m_k(\alpha_q) \tag{13}$$

$$pls_k(\alpha_i) = \sum_{r:\alpha_r \cap \alpha_i \neq \emptyset} m_k(\alpha_r) \tag{14}$$

In regard of this, data fusion can be performed to find the merged weight d_i respect to event α_i. Indeed, it is the same belief (see Equation 15) expression, such that $pls(\alpha_i) = 1 - bel(\overline{\alpha_i})$,

$$d_i = bel(\alpha_i) = m(\alpha_i) = \bigoplus_{k=1}^{K} m_k(\alpha_i) \tag{15}$$

where each piece of evidence is blended by making use of Dempster's rule of combination, as follows:

$$m_1(\alpha_i) \bigoplus m_2(\alpha_i) = \frac{\sum_{q,r:\alpha_q \cap \alpha_r = \alpha_i} m_1(\alpha_q) m_2(\alpha_r)}{1 - \sum_{q,r:\alpha_q \cap \alpha_r = \emptyset} m_1(\alpha_q) m_2(\alpha_r)} \tag{16}$$

Summing up, Figure 5 depicts system blocks for DCTE framework.

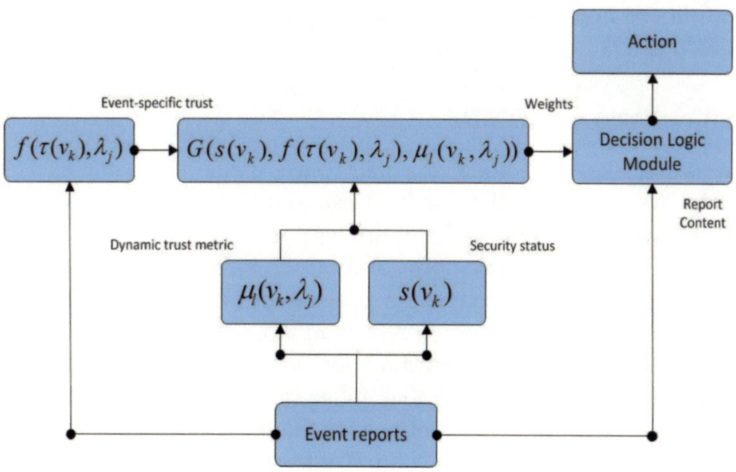

Figure 5. Data-centric Trust Establishment system blocks [21].

3.3.4. DCTE remarks

DCTE model appeals for a simple, system- and protocol-independent framework, composed basically by two stages: trustworthiness assessment and decision logic. Not only, the trust model is mathematically described, but also the entire VANET environment. Therefore, composite variables translated into vehicles typification, event-reports, system tasks, security issues and evaluation metrics are integrally controlled by modules of general description. The intercourse of functions is based on plain operations, whilst outputs remain within unit-value intervals. Furthermore, the induction of Bayesian inference and Dempster-Shafer theory as fusion techniques offers a novel view for handling of unions and intersections of compound events. So far, CoRS and TRIP models performed either polling or weighting procedures to cope with that issue, but formal results are not sufficiently derived.

As the foregoing fellow models, DCTE delegates the authentication, integrity and confidentiality issues on exogenous mechanisms; ergo nodes identities and credentials are supposed to be distributed beforehand. And then, the exposure to security and privacy-related attacks are current concerns. Moving apart, one of the premises adopted by DCTE is the ephemeral nature of nodes' relationships in VANETs, which urges to perform trustworthiness assessment per event. The actual impact of such a situation on information sparsity needs further research.

3.4. Distributed Emergent Cooperation through Adaptive Evolution - DECADE

In [24], a game theoretic trust model is proposed, whose aim is encouraging forwarding cooperation in MANETs. The model, called DECADE, uses a non-cooperative game to achieve cooperation among rational nodes and isolation of selfish nodes. Furthermore, due to the distributed nature of the evolution algorithm and the trust evaluation mechanisms, forwarding cooperation emerges with a low overhead on computational and communication resources. Although DECADE was designed for general MANET environments, there are several elements of interest in DECADE that can be considered specifically in VANETs. So, in this section we briefly describe DECADE principles and then discuss its potential contributions to VANETs.

3.4.1. The Basic Game Model in DECADE

In transmitting a packet, every node in a path plays a role as a source, intermediate, or destination node. As an intermediate node, it has to decide whether to forward or discard a received packet based on a strategy. This strategy will depend on the trust level that the intermediate node has on the source node, and on the recent behavior of the network as a whole with its own packets. Each node is supposed to be able to observe the decisions taken by its neighbors and by all nodes preceding it in a path, so the trust on each observed node can be computed as the number of forwarding decisions among the last m observations. The limited memory, m, accounts for the fact that nodes can change their strategies continuously in order to adapt their behavior to environmental changes, so its value obeys a trade-off: m has to be large enough to obtain a fair evaluation of the forwarding rate, but m has to be small enough to ensure that the forwarding rate actually corresponds to the current strategies.

An intermediate node must be very careful to discard a packet because observing nodes can reduce their trust on it; but it cannot cooperate indiscriminately in order not to become a "sucker", wasting valuable resources forwarding packets from selfish nodes. Taking this trade-off into account, the strategy in the forwarding game is encoded by a string of bits that represents the decision of discarding or cooperating, depending on the trust level that the node has in the source node, and on the number of successful transmissions within the last k own packets.

Each node participates in repeated games, where the decision to cooperate or discard of each intermediate node obeys to its current strategy. A game consists on the successful or failed transmission of a packet. Whenever a node is ready to send a packet, it chooses the most trusted path to the destination, i.e., the one that maximizes the probability that the packet gets its intended destination (which can be found through a shortest path routing algorithm under the appropriate distance metric). Then, the source node sends the packet on this path and each intermediate node decides whether to forward it or to discard it, according to its own strategies. The game ends either when the packet is delivered to its destination, or when an intermediate node decides to discard the packet. Once the game has finished, each intermediate node receives a payoff according to its decision, where forwarding decisions are paid in direct proportion to the trust level in the source node (rewarding cooperation), while discarding decisions are paid in inverse proportion (rewarding resource savings).

Given the game theoretic network trust model, it is important to find an optimal strategy for each node, so that the network as a whole maximizes both the cooperation among rational nodes and the isolation of selfish nodes. This will be done through a genetic algorithm that will evolve constantly to track the dynamical changes within the network.

3.4.2. Genetic Algorithm for Strategy Evolution

DECADE uses a distributed genetic algorithm of the cellular type with plasmid migration heuristics, which not only gives good results in term of the optimality of the converged solutions, but also exhibits good adaptability to changing conditions. Each node tries to maximize its payoff by exchanging periodically genetic information with its neighbors in order to evolve the strategy. As in a classical cellular genetic algorithm, each node receives the genetic information from all its one-hop neighbors, selects randomly two of them with a probability of being selected proportional to their fitness and, through the classical one point cross-over and mutation processes, combines them to construct a new strategy. This classical cellular mechanism is enhanced with a bacterial plasmid migration concept, where two heuristics are added. First, each node can accept or reject the new strategy depending on whether the reported fitness is greater or smaller than its own fitness. Second, each node can keep a copy of its best previous strategy so that, if during the current plasmid migration period the new strategy did not increase the fitness, the old strategy can be restored. An important heuristic in this evolution process is that, since each node keeps a record of its best strategy so far (plasmid genes instead of chromosomal genes), a node can replace the current strategy with the stored one, just before any strategy exchange among neighbors takes place. This heuristic enhances the exploratory capacity of the evolution process.

3.4.3. DECADE remarks

Intended to encourage forwarding cooperation in MANETs, DECADE achieves remarkable performance results in cooperation among rational nodes, isolation of selfish nodes and adaptability to changing environments. Furthermore, due to the distributed nature of the evolution algorithm and the trust evaluation mechanisms, cooperation emerges with a low overhead on computational and communication resources. However, although forwarding decisions could become an issue in some infotainment applications of vehicular ad hoc networks, the most important current problem in VANETs is in content trust and reputation, not addressed by DECADE. We just mentioned DECADE because it points out to the potential of complex systems engineering for trust and reputation systems in VANETs. Indeed, the whole system develops the cooperation as an emergent phenomenon, which appears as a consequence of individual decisions, based on local observations. Each node wants to save its scarce resources by using them rationally, seeking the cooperation of intermediate nodes to deliver their own packets. As a consequence of the local interactions among nodes, the global cooperative behavior arises, with the reported performance benefits. This approach should be explored more extensively in the engineering of VANET systems.

4. Trust and reputation model comparison

In regard of the foregoing trust and reputation passage, including general specifications, design criteria and structured mechanisms, we conclude the present chapter with an overall analytical comparison upon the studied models. Firstly, Table 3 gathers the capabilities of CoRS, TRIP, DCTE and DECADE with respect to the models' considerations, as a matter of design, implementation and support. Please note, three possible qualifications (T=Total, P=Partial, F=Fail) can be given in order to reflect their strengths and downsides.

Criterion	CoRS	TRIP	DCTE	DECADE
Low Complexity	P	P	T	T
Scalability	T	T	T	P
Sparsity	T	P	P	P
Security and privacy-related	T	P	P	P
Mobility patterns independent	P	P	P	P
Trust and reputation decentralization	P	P	T	T
Confidence measure	F	T	T	P
Event and spatiotemporal specification	T	T	T	T

Table 3. Criteria comparison among trust and reputation models.

Closing up, the qualification chart onto *low complexity* criterion. CoRS and TRIP protocols incorporates multi-purpose modules to improve the efficiency on trustworthiness computation, messages convergence, resilience and others. However, the employment of those metrics moving upward the protocol's performance, also impacts noxiously the altogether complexity levels. Likewise, *trust and reputation decentralization* criterion is partially fulfilled by CoRS and TRIP protocol, while DCTE and DECADE achieves a "Total" mark. One more time, the model conceptualization is intrinsically related to third-party entities, which are set up to enhance particular aspects in security matters. Indeed, DCTE and DECADE are also pending to explain how the confidentiality, integrity and authentication issues are natively handled. In the meantime, the top score is granted, assuming the engagement of exogenous mechanisms.

A limited performance seems to be equally exhibited by all the examined protocols with respect to *sparsity*. Taking CoRS out of this group; the introduction of cryptographic thresholds boosts the dissemination of information sources, according to the scarcity conditions on the road. From the bulky generation of shares, the model previously knows the actual capabilities of the mobile nodes to interact among them. Then, scenarios with lack of resources can be foreseen based on some degree of certainty. On the contrary, the remaining models do not follow a self-sustained strategy to adapt their mechanisms to selective information retrieval. However, potential potential bases are grounded to achieve flexible thresholds on trust and reputation assessment. For instance, TRIP and its fuzzy set of rules might be easily extended to actively manage to this requirement. A quite similar diagnosis can be emitted about *mobility patterns independent* criterion, where every protocol is marked as "Partial". Notwithstanding, the assurance of a "Total" mark on this concept would require a massive set of tests, regarding simulated and real scenarios. Until now, the achievement of common levels of agreement on this matter are very unlikely.

Outstandingly, *scalability* and *event and spatiotemporal specification* criteria are satisfactorily attained by the group of models. The former one can be possibly explained by the accelerated development of powerful processing platforms, whose physical resources are capable to adjust to the ongoing demands, neatly. In turn, replaying to what, when and where issues regarding the events on the road shall be imposed as a compulsory requirement for a trust and reputation model. Otherwise, practical usages of such protocols could be easily questioned.

Furthermore, trustworthiness computation metric is definitively one of the most interesting aspects to be differentiated in each trust and reputation model. Therefore, it is worthwhile to take a quick tour around the employed assessment techniques. Considering CoRS protocol, *majority voting* stands out as the selected method to obtain the trust level TL upon a particular event. Thus, Equation 17 depicts the value computed by K nodes, where each one contributes with a +1 reward for a confirmed on road event α_i. Likewise, negative or neutral values may apply in case of fake or deceiving information.

$$TL = \frac{1}{K} \sum_{k=1}^{K} f(\alpha_i) \tag{17}$$

In respect to TRIP protocol, *weighted polling* acts as the calculation method for trustworthiness modeling. Thereafter, all given recommendations or reports about an event $f(\alpha_i)$ is affected by an scalar weight w, whose value takes into consideration information sources, penalties and rewards history, VANET complexity, etc. The trust level TL performed by TRIP mode follows the general rule in Equation 18.

$$TL = \frac{1}{K} \sum_{k=1}^{K} w * f(\alpha_i) \tag{18}$$

For DCTE framework, a couple of data fusion techniques are designated for trust and reputation measurement. In section 3.3.2, we have introduced Bayesian inference as a statistical approach and Dempster-Shafer theory for evidence evaluation inspired in human reasoning. For the implicit mathematical description refer to the corresponding section.

5. Conclusions

Vehicular Ad Hoc Networks constitute an emergent and fascinating research field.Many conceptual architectures, abstract models and heuristic-derived systems are continuously proposed to cope with the many issues that arise in this area. In this chapter we have made an effort to describe four of the most remarkable approaches in literature.

CoRS model implements threshold cryptography in a distributed way, achieving low complexity in security management. In fact, CoRS might be considered one of the most efficient and effective protocols, although special attention should be given to crowded networks, since the distribution of repeated shares might lead to system's collapse. Similarly, the collator player is a potential faulty point because it is the only party designated to set time and location of events.

A strength of the TRIP protocol is the use of multiple information sources like direct peers, referred nodes and central entities. Also the well-defined system of penalties and rewards makes of TRIP a robust framework for trust and reputation determination. A differentiator factor of TRIP is its support of trust and reputation scoring system on adaptable probabilistic functions. However, TRIP has some reliability difficulties to solve.

DCTE strives to deal with trustworthiness assessment, decision logic and environment description. These aspects are integrated through fusion techniques, generating representations of trust and reputation scores that achieve both simplicity and efficiency. Unfortunately, DCTE has major issues in confidentiality, authentication and integrity aspects, which are assumed to be carried out by third party schemes.

We also introduced DECADE as a newcomer model, given its good performance within MANETs. The protocol encourages cooperation among rational nodes, isolating selfish nodes with high adaptability to changing environments. Useful concepts such as the emergence of a cooperative behavior from simple individual decisions based on local observations, with very low overhead, points out at the convenience of facing trust and reputation mechanisms in VANETs through the theory of complex systems. Although DECADE only addresses the MANET problem of node trust and reputation, its emergent approach could be used in the most urgent VANET problem of content trust and reputation.

Finally, considering eight criteria (complexity, scalability, sparsity, security, mobility dependence, decentralization, confidence measure and event specification), we compare the four selected approaches and notice that none of them satisfy all the criteria. However, with the exception of independence on mobility, all the different criteria are satisfied but at least one approach. A good research line would be to exploit the advantages of each proposal looking for a general framework where to put over solid basis the development of Distributed Trust and Reputation Mechanisms for Vehicular Ad-hoc Networks.

Author details

Marcela Mejia[1] and Ramiro Chaparro-Vargas[2]

1 Universidad Militar Nueva Granada, Bogotá, Colombia
2 RMIT University, Melbourne, Australia

References

[1] M. Gerlach and F. Friederici. Implementing trusted vehicular communications. In *Vehicular Technology Conference, 2009. VTC Spring 2009. IEEE 69th*, pages 1 –2, april 2009.

[2] P. Papadimitratos, L. Buttyan, T. Holczer, E. Schoch, J. Freudiger, M. Raya, Zhendong Ma, F. Kargl, A. Kung, and J.-P. Hubaux. Secure vehicular communication systems: design and architecture. *Communications Magazine, IEEE*, 46(11):100 –109, november 2008.

[3] P. Ardelean and P. Papadimitratos. Secure and privacy-enhancing vehicular communication: Demonstration of implementation and operation. In *Vehicular Technology Conference, 2008. VTC 2008-Fall. IEEE 68th*, pages 1 –2, sept. 2008.

[4] J. P. Hubaux P. Papadimitratos, V. Gligor. Securing vehicular communications - assumptions, requirements, and principles. november 2006.

[5] F. Kargl, P. Papadimitratos, L. Buttyan, M. Muter, E. Schoch, B. Wiedersheim, Ta-Vinh Thong, G. Calandriello, A. Held, A. Kung, and J.-P. Hubaux. Secure vehicular communication systems: implementation, performance, and research challenges. *Communications Magazine, IEEE*, 46(11):110 –118, november 2008.

[6] M. Raya, P. Papadimitratos, and J.-P. Hubaux. Securing vehicular communications. *Wireless Communications, IEEE*, 13(5):8 –15, october 2006.

[7] P. Papadimitratos, L. Buttyan, J. P. Hubaux, F. Kargla, A. Kung, and M. Raya. Architecture for secure and private vehicular communications. 2007.

[8] T. Leinmüller, L. Buttyan, J. P. Hubaux, F. Kargl, R. Kroh, P. Papadimitratos, M. Raya, and E. Schoch. Sevecom - secure vehicle communication. june 2006.

[9] L. Buttyan and J. P. Hubaux. Security and Cooperation in Wireless Networks. http://secowinet.epfl.ch, 2007. Cambridge University Press.

[10] P. Papadimitratos and J. P. Hubaux. Report on the secure vehicular communications: Results and challenges ahead workshop. april 2008.

[11] D. Djenouri, W. Soualhi, and E. Nekka. Vanet's mobility models and overtaking: An overview. In *Information and Communication Technologies: From Theory to Applications, 2008. ICTTA 2008. 3rd International Conference on*, pages 1 –6, april 2008.

[12] Félix Gómez Mármol and Gregorio Martínez Pérez. Trip, a trust and reputation infrastructure-based proposal for vehicular ad hoc networks. *Journal of Network and Computer Applications*, 35(3):934 – 941, 2012. <ce:title>Special Issue on Trusted Computing and Communications</ce:title>.

[13] Jie Zhang. A survey on trust management for vanets. In *Advanced Information Networking and Applications (AINA), 2011 IEEE International Conference on*, pages 105 –112, march 2011.

[14] A. Tajeddine, A. Kayssi, and A. Chehab. A privacy-preserving trust model for vanets. In *Computer and Information Technology (CIT), 2010 IEEE 10th International Conference on*, pages 832 –837, 29 2010-july 1 2010.

[15] R. A. Chaparro-Vargas. A security infrastructure for an in-vehicle middleware based on device profile for web services. M.sc. thesis, Munich, 2010.

[16] Marcela Mejia, Néstor Peña, José L. Muñoz, Oscar Esparza, and Marco A. Alzate. A game theoretic trust model for on-line distributed evolution of cooperation in manets. *Journal of Network and Computer Applications*, 34(1):39 – 51, 2011.

[17] C. S. Eichler. *Solutions for Scalable Communication and System Security in Vehicular Network Architectures*. Dissertation, Technische Universität München, Munich, 2009.

[18] Aifeng Wu, Jianqing Ma, and Shiyong Zhang. Rate: A rsu-aided scheme for data-centric trust establishment in vanets. In *Wireless Communications, Networking and Mobile Computing (WiCOM), 2011 7th International Conference on*, pages 1 –6, sept. 2011.

[19] Adi Shamir. How to share a secret. *Communications of the ACM*, 22(11):612–613, november 1979.

[20] Anand Patwardhan, Anupam Joshi, Tim Finin, and Yelena Yesha. A data intensive reputation management scheme for vehicular ad hoc networks. In *Mobile and Ubiquitous Systems - Workshops, 2006. 3rd Annual International Conference on*, pages 1 –8, july 2006.

[21] M. Raya, P. Papadimitratos, V.D. Gligor, and J.-P. Hubaux. On data-centric trust establishment in ephemeral ad hoc networks. In *INFOCOM 2008. The 27th Conference on Computer Communications. IEEE*, pages 1238 –1246, april 2008.

[22] Qing Ding, Xi Li, Ming Jiang, and XueHai Zhou. Reputation-based trust model in vehicular ad hoc networks. In *Wireless Communications and Signal Processing (WCSP), 2010 International Conference on*, pages 1 –6, oct. 2010.

[23] N.-W. Lo and Tsai H.-C. A reputation system for traffic safety event on vehicular ad hoc networks. page 10, 2009.

[24] Marcela Mejia, Néstor Peña, José L. Muñoz, Oscar Esparza, and Marco Alzate. Decade: Distributed emergent cooperation through adaptive evolution in mobile ad hoc networks. *Ad Hoc Networks*, 10(7):1379 – 1398, 2012.

Section 4

Advanced Applications and Services

Chapter 7

Accurate and Robust Localization in Harsh Environments Based on V2I Communication

Javier Prieto, Santiago Mazuelas, Alfonso Bahillo,
Patricia Fernández, Rubén M. Lorenzo and
Evaristo J. Abril

Additional information is available at the end of the chapter

http://dx.doi.org/10.5772/55263

1. Introduction

With the arrival of global navigation satellite systems (GNSS), in-car navigation has increasingly become an essential tool for the automotive industry. However, the performance of GNSS is compromised in harsh environments where there is not a line of sight (LOS) to satellites, e.g., tunnels, covered parking areas and dense urban canyons [1]. Hence, in-car navigation requires a localization technology that operates with robustness in such circumstances. The development of vehicular ad-hoc networks (VANETs) provides a promising platform to fulfill this requirement [2].

In VANETs, an on-board unit (OBU) inside the vehicle communicates with other OBUs or with stationary roadside units (RSUs), in vehicle-to-vehicle (V2V) and vehicle-to-infrastructure (V2I) communications, respectively [3]. Cooperation between OBUs can provide good position estimates in V2V communication [4-5]. However, the quick topology changes required by V2V approaches make V2I communication be the preferred option for in-car navigation in harsh environments [6]. In V2I communication, the position of an OBU (the target) can be estimated from range-related measurements taken on the radio-frequency signals transmitted to and from the RSUs (the anchors) [7]. However, the changeable and unpredictable characteristics of the wireless channel in harsh environments make multipath and non-line of sight (NLOS) propagation conditions be predominant [8-9]. Therefore, conventional positioning systems designed for tractable and static signal behavior cannot guarantee an adequate performance.

The position information extracted from the radio-frequency signals varies according to the type of measurement taken. Techniques based on time of arrival (TOA) [9-10] or received signal strength (RSS) [11-12] measurements obtain range-related information, whereas techniques based on angle of arrival (AOA) or time difference of arrival (TDOA) measurements extract information related to directions or difference of distances, respectively [13-14]. AOA and TDOA measurements entail significant costs of antenna-array integration or synchronizing devices. In this chapter, we focus on RSS and TOA measurements that can provide accurate localization with an appropriate complexity.[1]

Range or position estimation is an inference problem where the observations are the RSS and TOA measurements [15-16]. From a Bayesian perspective, determining the posterior distribution of ranges or positions from observations is the optimal approach [17-25]. Then, ranges or positions can be obtained by means of the maximum a posteriori (MAP) or the minimum mean square error (MMSE) estimators.

The optimality of the above mentioned methods depends on the fit between the model assumed for the relationship between measurements and ranges or positions (i.e., the likelihood function) and the actual behavior of the measurements. Tractable and static models for the likelihoods based on Gaussian distributions accurately explain the behavior of measurements only in open areas [26-28]. For harsh environments, several techniques have been developed to address the complex behavior of wireless signal metrics. In the TOA case, the NLOS bias causes range overestimation. Thus, a common procedure is to detect and remove NLOS measurements [29]; other techniques utilize prior knowledge about this NLOS error to subtract it and adjust the measurements to their LOS values [27,30]. In the RSS case, the performance depends on the estimation of the parameters that characterize the propagation channel at each time [12,26]. Certain approaches deal with the dynamic nature of RSS metric through fingerprinting or machine learning [11,21,31]; however, their accuracy is sensitive to fast environmental changes and they do not fuse different signal metrics.

Range and position estimation can be improved by exploiting the relationship among positions in time through Bayesian filtering. Kalman filtering techniques rely on Gaussian models that are not adequate for harsh environments. Different alternative methods based on variations of such filters, as well as on particle filters (PFs), have been proposed: low complexity nonlinear/non-parametric adaptive modeling is used for filtering of RSS fingerprints in [11,21]; recursive Bayesian estimation together with multipath and NLOS propagation effects are considered in [22-23]; TOA and RSS data fusion is performed in [32-34]; hybrid information is exploited by particle filtering in [24]; and RSS/TOA Bayesian fusion for multipath and NLOS mitigation are performed in [25]. However, these methods require prior information achieved by arduous training phases or rely on assumptions non-realistic for harsh environments, such as Gaussian and static models.

This chapter presents a framework for adaptive data fusion to handle the difficulties described above, based on non-parametric dynamic modeling of the likelihood. The subsequent usage of a PF leads to the adaptive likelihood particle (ALPA) filter. As we show, the estimation can be carried out without requiring any calibration stage, thus enabling localization capabilities

1 In the TOA case, measuring the round-trip time avoids the technical difficulty of time synchronization among the nodes.

http://dx.doi.org/10.5772/55263

to pre-existing wireless infrastructures, such as VANETs based on V2I communication. The main contributions of this chapter are as follows:

- We present techniques for adaptive and systematic modeling of the relationship between measurements and positions, by means of a dynamic and empirical likelihood function.
- We present a model for Bayesian fusion of TOA and RSS measurements, based on nonlinear and non-Gaussian Bayesian filtering and the likelihoods derived over time.
- We show the suitability of the proposed techniques by experimentation performed using common wireless local area network (WLAN) devices.
- We show the near-optimality of the method by comparing its performance to the posterior Cramér-Rao lower bound (CRLB).

Both empirical and simulation results show that the proposed methods significantly improve the accuracy of conventional approaches with an important reduction on the number of measurements needed.

The structure of the rest of this chapter is as follows: Section II defines the position estimation problem; Section III addresses this problem under a hidden Markov model (HMM) and defines the dynamic and measurements models; Section IV presents the adaptive data fusion technique for likelihood modeling and the recursive Bayesian approach for solving the resulting non-linear and non-Gaussian problem; Section V shows the experimental and simulation results; Section VI includes a discussion on complexity; and finally, Section VII draws the conclusions.

Notations: The notation $p(\mathbf{x})$ is the probability density function (pdf) of the random variable $\mathbf{x}$; $f^{(m)}(\bullet)$ denotes the mth derivative of a real function f evaluated in its argument; $f[k]$ for $k \in \mathsf{N}$ denotes the value of the function f evaluated in $t_k \in \mathsf{R}$; $\mathbf{X}[k]$ denotes the set $\{\mathbf{x}[i], i=1, \cdots, k\}$; if M is a positive integer, M^M denotes the M Cartesian power of $\{1, \ldots, M\}$; finally, $\bar{\mathbf{z}}$ denotes the sample mean of the components of a vector $\mathbf{z}$.

2. Problem statement

In the following, we consider a two-dimensional scenario where a mobile target (e.g., a car equipped with an OBU) moves freely. To determine its position, the target communicates with several anchors (the RSUs) with known positions. Since the localization system can get measurements in discrete times $\{t_k, k \in \mathsf{N}\}$, we are interested in estimating the sequence $\{\mathbf{x}[k], k \in \mathsf{N}\}$ from a sequence of measurements $\{\mathbf{z}[k], k \in \mathsf{N}\}$. The entries of vector $\mathbf{x}[k]$ can be the distances between the target and each anchor or the coordinates of the mobile target's position. The entries of vector $\mathbf{z}[k]$ are RSS and TOA measurements.

Next section establishes the probabilistic relationship between vectors $\mathbf{x}[k]$ and $\mathbf{z}[k]$ by modeling this problem as an HMM, and defining the state vector, $\mathbf{y}[k]$, that consists of vector $\mathbf{x}[k]$ and several of its derivatives.

3. Hidden Markov model

In addition to the information conveyed by the measurements, the fact that the sequence $\{\mathbf{x}[k],\ k\in\mathbb{N}\}$ is highly correlated in time can likewise be used as another source of information. The position of the target cannot change abruptly in a small lapse of time; hence, we can model the evolution in time of positions or Euclidean distances to each anchor as an analytic function. Being $x(t)$ a component of the position or the distance to an anchor, we can approximate its value in t_{k+1} by using the nth-order Taylor expansion in t_k,

$$x[k+1]\approx x[k]+x^{'}[k]\Delta t+x^{''}[k]\frac{\Delta t^2}{2}+\cdots+x^{(n)}[k]\frac{\Delta t^n}{n!} \tag{1}$$

where $\Delta t=(t_{k+1}-t_k)\in\mathbb{R}$ is the sampling interval. The error in this approximation is

$$x^{(n+1)}(\xi_0)\frac{\Delta t^{n+1}}{(n+1)!}$$

and $\xi_0\in\mathbb{R}$ is a point in the interval $[t_k,\ t_{k+1}]$. Therefore, the error in the approximation (1) depends directly on Δt, on the smoothness of $x(t)$ (represented by the $(n+1)$th derivative), and on the order of the approximation.

The correlation in time expressed in (1) implies that $\{\mathbf{x}[k],\ k\in\mathbb{N}\}$ is not a Markov chain, i.e., the current distance or position depends not only on the previous one. However, calling $\mathbf{y}[k]$ the positional-state consisting of the distance or the position and its n derivatives, we can assume that $\{\mathbf{y}[k],\ k\in\mathbb{N}\}$ is a Markov chain. Moreover, we can likewise assume that, conditioned on $\{\mathbf{y}[k],\ k\in\mathbb{N}\}$, $\{\mathbf{z}[k],\ k\in\mathbb{N}\}$ is a sequence of independent random variables, i.e., given the current positional-state vector, the measurements $\mathbf{z}[k]$ are independent of all previous and future positional-states and measurements [18]. These assumptions let to build an HMM in which the positional-state vectors $\{\mathbf{y}[k],\ k\in\mathbb{N}\}$ form a non-observable Markov chain, and what is available is the other stochastic process $\{\mathbf{z}[k],\ k\in\mathbb{N}\}$, linked to the Markov chain in that $\mathbf{y}[k]$ governs the distribution of $\mathbf{z}[k]$ [35] (see Figure 1).

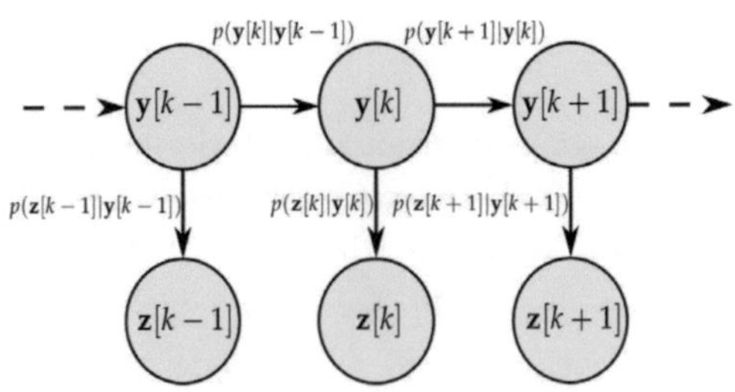

Figure 1. Hidden Markov Model for positional-states and measurements evolution. The relationship between $\mathbf{y}[k]$ and $\mathbf{y}[k-1]$ and the relationship between $\mathbf{z}[k]$ and $\mathbf{y}[k]$ are the only two kinds of dependence.

The conditional independence assumptions reflected in Figure 1 lead to two kinds of dependence between the random variables [36],

- *Dynamic model*: establishes the relationship between the state vector in time t_k and the state vector in time t_{k-1}, i.e., $p(\mathbf{y}[k] \mid \mathbf{y}[k-1])$.
- *Measurements model*: establishes the relationship between the measurements and the state vector in each time, i.e., $p(\mathbf{z}[k] \mid \mathbf{y}[k])$.

Then, the joint distribution of all the random variables involved in the process is given by,[2]

$$\begin{aligned} p(\mathbf{Y}[k],\mathbf{Z}[k]) &= p(\mathbf{y}[1])p(\mathbf{z}[1]\mid\mathbf{y}[1])\prod_{i=2}^{k} p(\mathbf{y}[i]\mid\mathbf{y}[i-1])p(\mathbf{z}[i]\mid\mathbf{y}[i]) \\ &= p(\mathbf{Y}[k-1],\mathbf{Z}[k-1])p(\mathbf{y}[k]\mid\mathbf{y}[k-1])p(\mathbf{z}[k]\mid\mathbf{y}[k]) \end{aligned} \tag{2}$$

The modeling as an HMM shown in (2) makes possible to infer the posterior distribution $p(\mathbf{y}[k] \mid \mathbf{Z}[k])$ through a recursive process. In the specific case where the dynamic and measurements model are linear-Gaussian, the posterior distribution is also Gaussian, and the Bayesian inference can be optimally performed by the celebrated Kalman filter (KF) [19]. In the following, we describe these models for harsh propagation environments, showing that the dynamic model can be assumed linear with a wide generality, whereas this assumption for the measurements model yields inaccurate performances.

3.1. Dynamic model

The dynamic model of the positional-state vector can be obtained from the evolution in time given by (1), and by approximating each mth derivative, for $m=1, \ldots, n$, by its $(n-m)$th-order Taylor expansion, as

$$\mathbf{y}[k+1]=\mathbf{F}_k\mathbf{y}[k]+\mathbf{n}_d[k] \tag{3}$$

where

$$\mathbf{F}_k=\begin{pmatrix} 1 & \Delta t & \frac{\Delta t^2}{2} & \cdots & \frac{\Delta t^n}{n!} \\ 0 & 1 & \Delta t & \cdots & \frac{\Delta t^{n-1}}{(n-1)!} \\ \vdots & \ddots & \ddots & \ddots & \vdots \\ 0 & \cdots & 0 & 1 & \Delta t \\ 0 & \cdots & \cdots & 0 & 1 \end{pmatrix} \tag{4}$$

is the transition matrix, and $\mathbf{n}_d[k]$ is the error in the approximations. For example, in the case of estimating a one-dimensional parameter, $x[k]$, the error $\mathbf{n}_d[k]$ is given by

2 This probabilistic model is a generalization of the maximum likelihood approach in which the estimation is accomplished for a given time instant, t_k, neither considering previous nor future positional-states and measurements. In this case, $p(\mathbf{y}[k], \mathbf{z}[k]) \propto p(\mathbf{z}[k] \mid \mathbf{y}[k])$

$$\mathbf{n}_d[k]=\begin{pmatrix} \frac{\Delta t^{n+1}}{(n+1)!} x^{(n+1)}(\xi_0) \\ \frac{\Delta t^{n}}{n!} x^{(n+1)}(\xi_1) \\ \vdots \\ \Delta t x^{(n+1)}(\xi_n) \end{pmatrix} \tag{5}$$

where $\xi_0, \ldots, \xi_n$ are values in the interval $[t_k, t_{k+1}]$. The values taken by the $(n+1)$ th derivative of $x(t)$ in the unknown points $\xi_0, \ldots, \xi_n$ are modeled as realizations of a random variable that can be assumed to be zero-mean Gaussian variable with a standard deviation $\sigma_{d^{(n+1)}}$ [18-19]. Then, we can model the evolution in time of $x[k]$ as a random walk. Therefore, the dynamic model is a discrete Wiener process velocity (DWPV) model or a discrete Wiener process acceleration (DWPA) model if we use the second- or third-order Taylor expansion, respectively [18]. Hence, the dynamic model, $p(\mathbf{y}[k] \mid \mathbf{y}[k-1])$, can be assumed linear-Gaussian.

3.2. Measurements model

The second ingredient to characterize the HMM is the measurements model or likelihood, $p(\mathbf{z}[k] \mid \mathbf{y}[k])$. This probability distribution relates the measurements to the positional-state. In the case of range-related measurements, we have that $p(\mathbf{z}[k] \mid \mathbf{y}[k]) = p(\mathbf{z}[k] \mid d[k])$, irrespectively of the positional-state used. In the following, we describe realistic models for the relationship between distances and RSS/TOA measurements in concordance with previous essays [10,12].

3.2.1. RSS measurements

In a given specific instant and place, the RSS values are affected by the distance between emitter and receiver. The attenuation caused by the distance between two nodes is known as *path-loss* and is proportional to this distance raised to a certain exponent, called *path-loss exponent* [7,12,15,26]. However, the RSS values are likewise affected by a wide range of unpredictable factors, such as multipath propagation (fast fading) and shadowing (slow fading) [37]. By reflecting these factors in the Friis transmission equation for free-space, the relationship between the received signal strength, P_r, and the distance, $d[k]$, is given by [26],

$$P_r = \frac{G_t G_r}{4\pi} P_t \frac{g^2 \gamma}{(d[k])^{\beta_s}} \tag{6}$$

where P_t is the transmitted power, G_t and G_r are the transmitter and receiver gains, respectively, g and γ are the parameters of the Rayleigh/Rician and log-normal distributions that model the fast and slow fading, respectively, and β_s is the path-loss exponent corresponding to the specific propagation environment [37].

By following the procedure described in [26] and taking logarithmic units, we obtain the measurements model for RSS values,

$$z_s[k] = \alpha_s - 10\beta_s \log_{10}(d[k]) + n_s[k] \tag{7}$$

where $z_s[k] \in \mathrm{R}$ is the RSS measured value and α_s a constant that depends on P_t, G_t, G_r and the fast and slow fading [12,15,26]. Finally, $n_s[k]$ is a noise term caused by shadowing that has zero mean in cases where the parameters α_s and β_s fit perfectly the current propagation conditions [12,15,26]. In practice, the value of α_s can be previously known [26]. However, in realistic scenarios, the path-loss exponent, β_s, used to relate RSS values to distances, does not fit exactly the actual propagation conditions [12], and hence, the noise term, $n_s[k]$, will have a non-zero mean proportional to the logarithm of the distance.

3.2.2. TOA Measurements

The distance between emitter and receiver also affects the time taken by the signal to be propagated from one node to the other. By assuming known the signal speed, we can infer this distance by means of a linear transformation of the TOA values. Due to the technical difficulty of synchronizing devices in a wireless network, techniques that use round-trip time estimation are the most attractive to estimate delays [10,28]. In this case, the processing time at the device that has to transmit the echo causes the relationship between TOA and distance to be affine linear (it has an intercept term). Then, we can model the relationship between the delay, $z_\tau[k]$, measured at time t_k, and the distance at that time, $d[k]$, as,

$$z_\tau[k] = \alpha_\tau + \beta_\tau d[k] + n_\tau[k] \tag{8}$$

where α_τ and β_τ are constants that can be estimated by a linear regression of measurements previously obtained [28,38-39]. The term $n_\tau[k]$ models the noise that is ussually assumed to be zero-mean and Gaussian in case of LOS propagation. However, in case of NLOS propagation, it is currently not known how to accurately model such error term, where several statistical distributions taking positive values, such as Exponential, Rayleigh, Weibull or Gamma, have been used in the literature [26-28].

From the above discussion, we can notice that in all cases the expected value of the measurements is $\mathrm{E}\{z\} = f(d[k]) + b$, where f is a linear or logarithmic function, and b is a systematic error in the model. In addition, we can point out that in harsh environments:

1. the relationship between measurements related to distances and distances is nonlinear and non-Gaussian;

2. such relationship highly depends on the propagation environment that can change rapidly.

These two factors render the linear-Gaussian assumption inadequate for the measurements model, $p(\mathbf{z}[k] \mid \mathbf{y}[k])$, in harsh environments. Therefore, common inference techniques that use naive and static models may obtain poor results in realistic dense cluttered scenarios.

4. Bayesian adaptive RSS/TOA fusion

Conventional non-Bayesian approaches for parameter estimation are based on maximum-likelihood (ML) estimation (in our case the maximization of $p(\mathbf{z}[k] \mid \mathbf{y}[k])$). ML commonly assumes tractable models for the likelihood (e.g., Gaussian likelihoods yield a least squares problem), while more intricate models are usually solved by means of expectation-maximization (EM) algorithm [40-41]. In the event that certain prior information about the parameter of interest is available, we can achieve a better estimator by adding this new information. If this prior information is the correlation in time of positional-states, it can be exploited through sequential Bayesian inference. In the following, we briefly describe such estimation process and present the adaptive likelihood particle (ALPA) filter for Bayesian inference based on RSS and TOA non-parametric adaptive likelihoods.

4.1. Bayesian inference

In the above mentioned context, the task is to determine the posterior distribution of positional-states given the measurements, $\mathbf{Z}[k]$, from the knowledge of the prior, $p(\mathbf{y}[k])$, and the likelihood, $p(\mathbf{z}[k] \mid \mathbf{y}[k])$, by using the Bayes' rule [19,42]. The knowledge about the prior distribution, $p(\mathbf{y}[k])$, can come from several avenues, e.g., from environmental knowledge. In this chapter, we use as prior knowledge the positional-states inferred in previous instants over the framework offered by the HMM above explained (see Figure 1). However, any other kind of prior information can be incorporated analogously.

In the case of modeling the positional-state and measurements evolution as an HMM, the expression (2) provides a way to determine the posterior distribution iteratively,

$$p(\mathbf{Y}[1] \mid \mathbf{Z}[1]) = \frac{p(\mathbf{y}[1], \mathbf{z}[1])}{p(\mathbf{z}[1])} = \frac{p(\mathbf{y}[1])p(\mathbf{z}[1] \mid \mathbf{y}[1])}{p(\mathbf{z}[1])}$$

and for $k > 1$,

$$\begin{aligned} p(\mathbf{Y}[k] \mid \mathbf{Z}[k]) &= \frac{p(\mathbf{Y}[\mathrm{k}], \mathbf{Z}[\mathrm{k}])}{p(\mathbf{Z}[\mathrm{k}])} \\ &= \frac{p(\mathbf{z}[k] \mid \mathbf{y}[k])p(\mathbf{y}[k] \mid \mathbf{y}[k-1])p(\mathbf{Y}[\mathrm{k}-1], \mathbf{Z}[\mathrm{k}-1])}{p(\mathbf{Z}[\mathrm{k}])} \\ &= \frac{p(\mathbf{z}[k] \mid \mathbf{y}[k])p(\mathbf{y}[k] \mid \mathbf{y}[k-1])p(\mathbf{Y}[\mathrm{k}-1] \mid \mathbf{Z}[\mathrm{k}-1])}{p(\mathbf{z}[\mathrm{k}] \mid \mathbf{z}[\mathrm{k}-1])} \end{aligned} \tag{9}$$

From the posterior distribution, $p(\mathbf{Y}[k] \mid \mathbf{Z}[k])$, we can estimate $\mathbf{y}[k]$ by,

$$p(\mathbf{y}[k] \mid \mathbf{Z}[k]) = \int p(\mathbf{Y}[k] \mid \mathbf{Z}[k]) d\mathbf{Y}[k-1] \tag{10}$$

leading to a process called filtering.[3] By replacing (9) in (10) we obtain,

$$p(\mathbf{y}[k] \mid \mathbf{Z}[k]) = \frac{p(\mathbf{z}[k] \mid \mathbf{y}[k]) \int p(\mathbf{y}[k] \mid \mathbf{y}[k-1]) p(\mathbf{Y}[k-1] \mid \mathbf{Z}[k-1]) d\mathbf{Y}[k-1]}{p(\mathbf{z}[k] \mid \mathbf{z}[k-1])} \tag{11}$$

By assuming known the posterior distribution at t_{k-1}, $p(\mathbf{Y}[k-1] \mid \mathbf{Z}[k-1])$, we can perform the filtering process in two steps [19],

1. *Prediction*: from the dynamic model we obtain the prediction of the positional-state in time t_k, given the measurements until time t_{k-1},

$$p(\mathbf{y}[k] \mid \mathbf{Z}[k-1]) = \int p(\mathbf{y}[k] \mid \mathbf{y}[k-1]) p(\mathbf{Y}[k-1] \mid \mathbf{Z}[k-1]) d\mathbf{Y}[k-1] \tag{12}$$

2. *Update*: from the measurements model we correct the prediction when a new set of measurements, $\mathbf{z}[k]$, is available in time t_k,

$$p(\mathbf{y}[k] \mid \mathbf{Z}[k]) = \frac{p(\mathbf{z}[k] \mid \mathbf{y}[k]) p(\mathbf{y}[k] \mid \mathbf{Z}[k-1])}{p(\mathbf{z}[k] \mid \mathbf{z}[k-1])} \tag{13}$$

and the normalization constant,

$$p(\mathbf{z}[k] \mid \mathbf{z}[k-1]) = \int p(\mathbf{z}[k] \mid \mathbf{y}[k]) p(\mathbf{y}[k] \mid \mathbf{Z}[k-1]) d\mathbf{y}[k] \tag{14}$$

Hence, the objective is to infer the hidden positional-state vector in each time, $\mathbf{y}[k]$, by using the information achieved by the measurements and the relationship between the variables in time. The Bayesian recursive process given by (12) and (13) avoids the need of reprocessing all the stored data since the posterior distributions are obtained iteratively. Figure 2 graphically explains the evolution of the distributions involved in the filtering process, for the problem of estimating the range between the OBU and an RSU, and for the problem of estimating the position of the OBU when it communicates with three RSUs.

In order to perform the described filtering process, we need the likelihood function of the measurements $p(\mathbf{z}[k] \mid \mathbf{y}[k])$. This function is a priori unknown in harsh environments, since the distribution of the error term in the measurements model is highly environmental-dependent and varies rapidly with time. In the RSS case, although the error term is usually assumed to be zero-mean Gaussian distributed, this assumption is too naive in realistic scenarios where, for example, only one estimation of the path-loss exponent is available [12,26]. For TOA measurements, this error term has been modeled with several parametric distributions such as Gaussian, Exponential, Gamma or Rayleigh [26-27,43] or by means of specific distributions obtained in each particular propagation environment [28,44]. In the following sections, we propose an adaptive likelihood function for data fusion that dynamically adjusts to the changing propagation conditions from the nature of the measurements collected in real time.

3 The positional-state $\mathbf{y}[k]$ can likewise be estimated by using the measurements until time t_{k+l}, leading to a process called smoothing if $l>0$ or prediction if $l<0$.

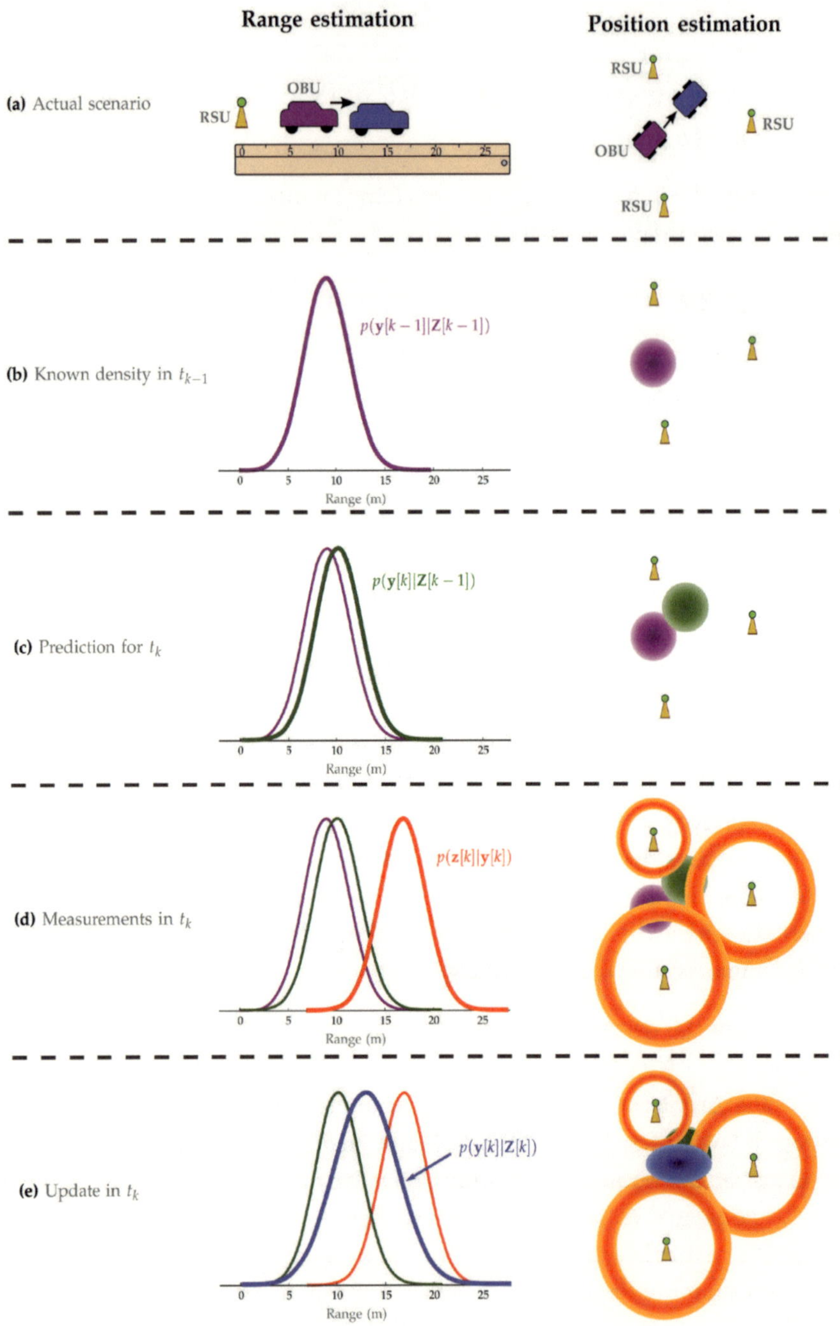

Figure 2. Density functions involved in filtering process for range and position estimation (darker zones have higher probability): (a) the target with the OBU moves in t_k with respect to its position in t_{k-1}; (b) the posterior density in t_{k-1} is known; (c) from the dynamic model we perform the prediction; (d) in t_k the target receives a new set of measurements; (e) from the likelihood we update the prediction to obtain the posterior density in t_k.

4.2. Adaptive likelihood for RSS/TOA fusion

The sets of RSS and TOA measurements obtained in each instant consist of samples from the random variable $z_s[k]$ and $z_\tau[k]$, respectively. As we show below, it is possible to represent the likelihood function in each instant and environment by using the set of samples through a non-parametric representation based on kernels [11,45-46].[4] After the reception of M RSS or M TOA measurements $\{z_k^i,\ i=1,\ \ldots,\ M\}$, we can approximate the pdf of $z_s[k]$ or $z_\tau[k]$ as

$$p\left(z\right) \approx \frac{1}{Mh}\sum_{i=1}^{M} K\left(\frac{z - z_k^i}{h}\right) \tag{15}$$

where $K(\bullet)$ is the kernel function and h is a positive number called bandwidth [11,45-46]. Several functions can be chosen for the kernel, where the most common is to use the standard Gaussian kernel [47], i.e.,

$$K(x) = \frac{1}{\sqrt{2\pi}} e^{-\frac{1}{2}x^2} \tag{16}$$

By assuming that the distribution of the measurements z, has the expression (15) in time t_k, we can obtain the likelihood relating distances to measurements in each instant k as the following result shows.

Proposition 1. Let $\mathbf{z}[k] = \{z_k^i,\ i=1,\ \ldots,\ M\}$ be a set of measurements (samples of z) related to the distance $d[k]$ by a model $\mathrm{E}\{z\} = \underline{f}(d[k]) + b$. Then, assuming z follows the distribution given by (15), and calling $\varsigma_{i,j} = z_k^j - z_k^i + \mathbf{z}[\mathrm{k}]$, the likelihood function of the measurements is

$$p(\mathbf{z}[k] \mid d[k]) = \frac{1}{(2\pi)^{M/2}(Mh)^M} \sum_{(i_1,\ldots,i_M)\in \mathrm{M}^M} \mathrm{E}_\mathrm{b}\left\{\exp\left(\frac{-1}{2h^2}\sum_{j=1}^{M}\left(\varsigma_{i,j} - f\left(d[k]\right) - b\right)^2\right)\right\} \tag{17}$$

where the expectation $\mathrm{E}_\mathrm{b}\{\bullet\}$ is taken with respect to the systematic errors, b, in the model.

Proof: see [48].

The Proposition 1 enables to obtain individual likelihoods from a set of measurements. Data fusion from different signal metrics (i.e., RSS and TOA) is carried out by combining these likelihoods. Let $\mathbf{z}_s[k]$ and $\mathbf{z}_\tau[k]$ be sets of RSS and TOA measurements, respectively, forming the set of measurements obtained in the instant k. Then, assuming that, given the real distance, $d[k]$, $\mathbf{z}_s[k]$ and $\mathbf{z}_\tau[k]$ are independent, we have that,

$$p(\mathbf{z}[k] \mid d[k]) = p\left(\mathbf{z}_s[k] \mid d[k]\right)p\left(\mathbf{z}_\tau[k] \mid d[k]\right) \tag{18}$$

where the likelihood of each kind of measurement can be dynamically obtained from (17).

---

4 A kernel function is a symmetric function (not necessarily positive) whose integral over the entire space is equal to one.

In order to describe how the presented adaptive data fusion operates, Figures 3-4 show the histogram of 100 RSS and 100 TOA measurements taken at a fixed distance with the measuring systems described in [12] and [10], respectively. These figures also represent the corresponding Gaussian pdf and the adaptive pdf obtained by means of the kernel-based expression given by (15) and (16).[5] From those figures, we can point out that, despite the fact that the true density is unknown, the presented adaptive pdf can express the dynamic behavior of RSS/TOA measurements in harsh environments with better accuracy than histogram and Gaussian density estimates [49-50].

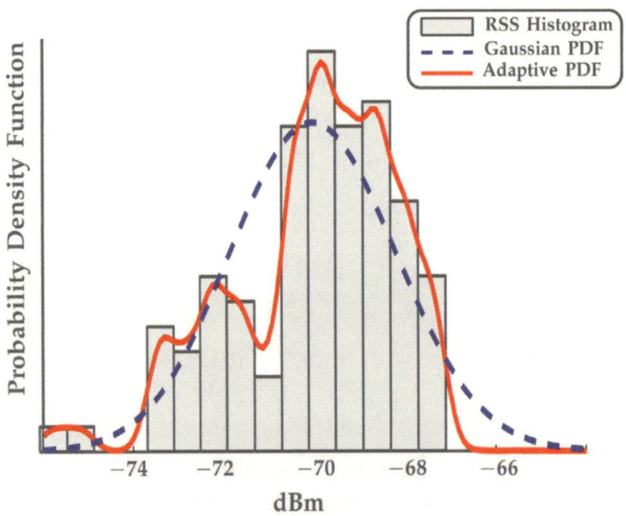

Figure 3. The adaptive density accurately approximates the complex randomness of RSS measurements in harsh environments.

5 In Figures 3-4 and in the following, we use a fixed bandwidth of one-half of the resolution of the measuring system [10, 12]. This election avoids both undersmoothed curves with too much spurious data artifacts, and oversmoothed densities that obscure the underlying nature of the measurements [46].

http://dx.doi.org/10.5772/55263

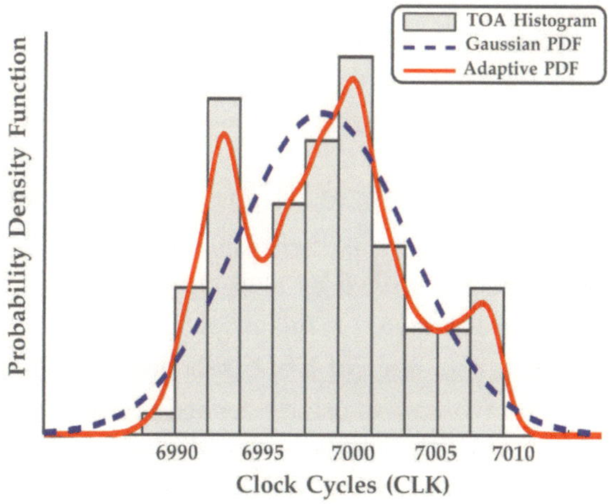

Figure 4. The adaptive density accurately approximates the complex randomness of TOA measurements in harsh environments.

In Figure 5 we illustrate the RSS/TOA data fusion process by representing the adaptive likelihood function obtained by means of expressions (17) and (18).[6]

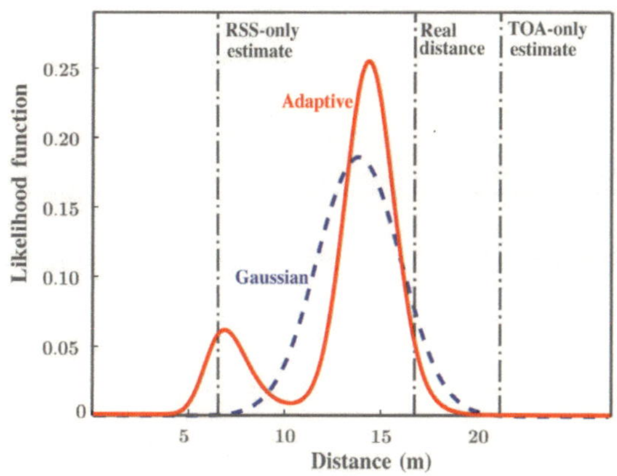

Figure 5. The adaptive RSS/TOA data fusion, defined by Proposition 1 and (18), results, in this case, in an improvement of 0.5 meters in ML estimator compared with the Gaussian case, which is equivalent to a reduction of 18% of the error.

From Figure 5, we can point out that the adaptive likelihood function provides more information about the distance than the Gaussian model, by combining the individual adaptive likelihoods obtained with RSS and TOA measurements. Moreover, the height of both functions reflects the more reliable information obtained by adaptive estimation. From that figure, we

6 In Figure 5 and in the following sections, we use coarse models for the measurements biases in accordance with previous essays [10,12]. Specifically, the RSS bias is modeled as a Gaussian $\mathrm{N}(0, \sigma_s)$ with σ_s=3 dBm, and the TOA bias as a Uniform distribution $\mathrm{U}(0, \gamma_\tau)$ with γ_τ=4 clock cycles.

also observe the improvement achieved by means of data fusion with respect to the individual estimates. This likelihood function leads to the ALPA filter defined in the following section.

4.3. Adaptive likelihood particle filter

Within the framework provided by the HMM, if both dynamic and measurements models are linear-Gaussian, all the posterior distributions are also Gaussian. In this case, all the involved density functions are completely described by their mean vectors and covariance matrices, obtained by a KF [19]. In the case of interest in this chapter, the models in the HMM are neither linear nor Gaussian, and then, the usage of KFs is suboptimal. In order to circumvent this drawback, the classical solution consists of using extended KFs (EKF) [23,25]. However, better performances can be obtained by PFs that let the usage of more general and flexible models [17,19] as the adaptive likelihood described in the previous section.

A PF represents the posterior distribution through a discrete distribution, where the support points and their probabilities are called particles and weights, respectively. To estimate the posterior distribution, we need to iteratively obtain a certain number of samples (particles) and probabilities (weights) capable of representing the posterior distribution. These particles and weights can be obtained by a method known as sequential-importance-sampling (SIS) [19,51], where the weight of the different particles can be determined by evaluating the likelihood function pointwise. Therefore, more realistic models such as the presented adaptive likelihood function for data fusion can be used, leading to the ALPA filtering algorithm describe in Table 1.

i. Initialization:

• Initial particles: draw N samples $\{\mathbf{y}_1^i,\ i=1,\ \ldots,\ N\}$ from the known density function $p(\mathbf{y}[1])$.

• Initial weights: $\omega_1^i = \frac{1}{N},\ i=1,\ \ldots,\ N$.

ii. Recursive estimation: for $k>1$,

• Particles in instant k from particles in instant $k-1$: draw N samples $\{\mathbf{y}_k^i,\ i=1,\ \ldots,\ N\}$ from the proposal distribution $q(\mathbf{y}[k]\,|\,\mathbf{y}_{k-1}^i,\ \mathbf{z}[k])$.

• From RSS measurements and Proposition 1, evaluate the weight of each particle. For $i=1,\ \ldots,\ N$

$$\tilde{\omega}_S^i = p(\mathbf{z}_S[k]\,|\,\mathbf{y}_k^i)$$

• From TOA measurements and Proposition 1, evaluate the weight of each particle. For $i=1,\ \ldots,\ N$

$$\tilde{\omega}_T^i = p(\mathbf{z}_T[k]\,|\,\mathbf{y}_k^i)$$

• Evaluate for $i=1,\ \ldots,\ N$

$$\tilde{\omega}_k^i = \omega_{k-1}^i \frac{\tilde{\omega}_S^i \tilde{\omega}_T^i p(\mathbf{y}_k^i\,|\,\mathbf{y}_{k-1}^i)}{q(\mathbf{y}[k]\,|\,\mathbf{y}_{k-1}^i,\ \mathbf{z}[k])}$$

• Normalization: for $i=1,\ \ldots,\ N$, compute

$$\omega_k^i = \frac{\tilde{\omega}_k^i}{\sum_{j=1}^{N} \tilde{\omega}_k^j}$$

Table 1. ALPA filtering.

To implement the algorithm detailed in Table 1, we have to choose a proposal distribution, where the most popular choice is to use the transition prior given by the dynamic model, i.e., $p(\mathbf{y}[k] \mid \mathbf{y}[k-1])$ [19]. This election leads to a rather simple expression for the weights,

$$\tilde{\omega}_k^i = \omega_{k-1}^i \tilde{\omega}_s^i \tilde{\omega}_\tau^i \tag{19}$$

Therefore, in order to use this algorithm, we have to obtain samples from the transition prior and evaluate the adaptive likelihood function pointwise. Figure 6 summarizes how this filter works with the proposal distribution chosen. First, we generate particles from the proposal distribution, in this case, the prior distribution, $p(\mathbf{y}[k] \mid \mathbf{y}[k-1])$, and then, their weights are updated according to the likelihood function, $p(\mathbf{z}[k] \mid \mathbf{y}[k])$. If the support of the proposal distribution does not cover the support of the likelihood function, only few particles will be in the *region of importance*, thus, the number of particles has to be increased in order to correctly approximate the posterior distribution.

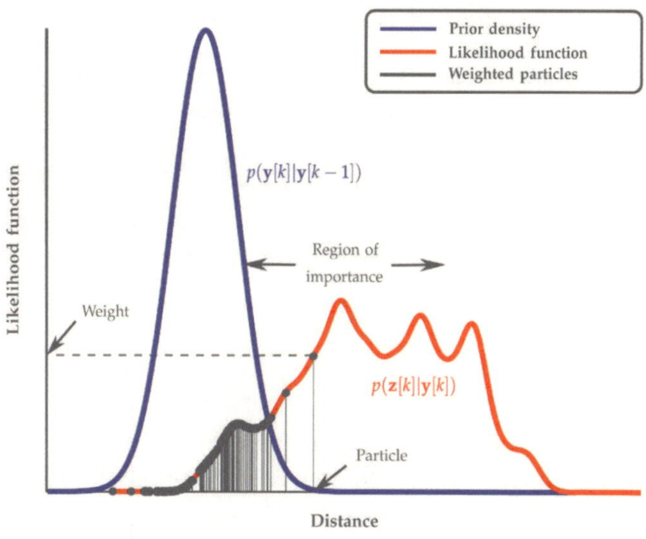

Figure 6. Transition prior and likelihood functions. Particles are obtained by sampling from the prior and weighting from the likelihood.

In this SIS algorithm, as k increases, the variance of the weights ω_k^i also increases, and therefore, after a certain number of steps, all but one particle will have negligible normalized weights. This problem is known as degeneracy [19]. To overcome this drawback, it is mandatory to perform a resampling step when a severe degeneracy is detected. A measure of degeneracy is the effective sample size N_{eff}, estimated as,

$$\hat{N}_{eff} = \frac{1}{\sum_{i=1}^{N} (\omega_k^i)^2} \tag{20}$$

where a small $\hat{N}_{eff}$ indicates a severe degenerancy. Therefore, when degenerancy is detected, N samples with uniform weights are drawn from the discrete representation of the posterior, given by the previous particles and weights, yielding a variant of SIS algorithm called sampling-importance-resampling (SIR) algorithm [19,52].

5. Results

The goal of this section is to quantify the performance of the methods presented in the above sections, leading to the ALPA filter. In order to do that, we obtained experimental data in a real indoor scenario by using the systems described in [10] and [12], and we ran numerous Monte Carlo simulations. In the following, we compare the performance of the introduced techniques with conventional approaches as well as with the CRLB.

We use the dynamic and measurements models above described together with the following state vector and prior information, depending on whether we estimate ranges or positions,

- *Range estimation*: we use a state vector $\mathbf{y}[k]=\left(d[k],\ d'[k],\ d''[k]\right)$. The standard deviation $\sigma_{d^{(3)}}$ is 1 m/s^3, which is roughly 50% of the maximum [18]. Furthermore, we add prior information about first and second derivatives of the distance, by considering they are distributed as Gaussians $\mathrm{N}\left(0,\ \sigma_{d'}\right)$ and $\mathrm{N}\left(0,\ \sigma_{d''}\right)$ respectively, where $\sigma_{d'}$=0.5 m/s and $\sigma_{d''}$=0.5 m/s^2.

- *Position estimation*: we use a state vector $\mathbf{y}[k]=(\mathbf{x}[k],\ \mathbf{v}[k],\ \mathbf{a}[k])$, where $\mathbf{x}[k]$ consists of the two-dimensional coordinates of the mobile target's position, and $\mathbf{v}[k]$ and $\mathbf{a}[k]$ are the velocity and the acceleration vectors. The same previous values for the deviations of the derivatives of the coordinates are used for dynamic and prior information.

For the experimental data, the target carried a laptop equipped with an IEEE 802.11b/g adapter and the measuring systems described in [10] and [12]. The anchors consisted of IEEE 802.11b/g access points (APs). In the RSS case, the anchors periodically sent beacon frames (at a frequency of MHz) and the RSS values were obtained based on the RSS indicator at target's adapter [12]. In the TOA case, the mobile target periodically sent request-to-send frames to each anchor (at a frequency of MHz), and a counter connected to the WLAN adapter saved the clock-cycles elapsed between the request and the reception of the corresponding clear-to-send frame [10]. For the results presented in this section, we refer as fusion the results of combining RSS and TOA data at every time-step.

5.1. Experimental results

As mentioned above, in a realistic scenario, NLOS propagation together with multipath effects constitute the major drawback of localization in harsh environments. This section illustrates the behavior of the proposed algorithm during a typical path followed by a mobile target in an indoor scenario. We carried out a measurement campaign inside an office building cluttered with clusters of objects and people moving freely in the area of the measurements. The propagation conditions were even harsher than the ones commonly find by an OBU placed within a car. Figure 8 shows the trajectory of 65 meters as well as the position of the 4 APs. It took 100 seconds to complete the whole trajectory, receiving a new set of measurements every second ($\Delta t = 1$s) from all the APs. As reflected in Figure 8, NLOS was always present when measuring with respect to AP3 and AP4, and only in a small percentage of positions there was a LOS between target and anchors AP1 and AP2.

In Table 2, we compare the error achieved with the proposed ALPA range estimation method in the presented scenario to the error obtained with conventional approaches [15,24]. We specify the results for RSS-only and TOA-only cases, and for their fusion. Specifically, we call,

- ML-RSS, ML-TOA, ML-Fusion: the range estimates obtained by means of the ML estimator. We utilize as likelihood function the convolution of the likelihood reported by the measurements (log-normal in the RSS case and Gaussian in the TOA case) and a Gaussian distribution corresponding to the bias.[7] The likelihood for the fusion is computed from (18).
- AML-RSS, AML-TOA, AML-Fusion: the ranges that correspond to the result of obtaining the maximum of the adaptive likelihood computed by means of Proposition 1, and (18) in the fusion case.
- EKF-RSS, KF-TOA, EKF-Fusion: the result of applying EKF and KF filters for RSS and TOA measurements, respectively, using the same bias distributions as in the ML case, and the dynamic model given by (3).
- ALPA-RSS, ALPA-TOA, ALPA-Fusion: the range estimates obtained by the ALPA filtering described in Table 1, where $N = 10\,000$ is the number of particles used.

We summarize for all these methods the quartiles of the absolute error in range estimates as well as the root mean squared error (RMSE), which incorporates both systematic (bias) and random errors. In order to study the influence of the number of measurements, M, in the final performance, all these statistics are shown for four different values.

Figure 7 depicts the pdf of the absolute error in range estimation after applying AML-Fusion and ALPA-Fusion methods, taking 10 RSS and 10 TOA measurements in each one of the positions of the target with respect to the four APs. Figure 7 likewise includes the ML-Fusion and EKF-Fusion methods in order to compare their behavior. Using only 10 measurements, ML-, AML-, EKF- and ALPA-Fusion obtain an error in range estimation lower than 3 meters

7 In order to guarantee a fair comparison, in Table 2 and in the following experiments, we select the values for the biases in accordance to the ones selected in Section 4. In this way, the RSS bias is modeled as a Gaussian $\mathrm{N}(0, \sigma_s)$ with $\sigma_s = 3$ dBm, and the TOA bias as a Gaussian $\mathrm{N}(\gamma_\tau/2, \gamma_\tau/4)$ with $\gamma_\tau = 4$ clock cycles.

for 55%, 65%, 73%, and 80% of the positions, respectively, which reflects the remarkable performance of the proposed algorithm.

	$M=5$		$M=10$		$M=50$		$M=100$	
	Quartiles	**RMSE**	**Quartiles**	**RMSE**	**Quartiles**	**RMSE**	**Quartiles**	**RMSE**
ML-RSS	1.64-3.12-5.45	7.01	1.28-2.94-4.96	5.32	1.36-2.72-4.68	4.34	1.27-2.74-4.74	4.58
ML-TOA	2.09-3.92-7.68	6.42	1.55-3.40-5.64	5.00	1.26-2.69-4.40	3.87	1.12-2.44-4.00	3.55
ML-Fusion	1.52-3.16-5.87	5.25	1.26-2.66-4.73	4.23	1.09-2.24-3.89	3.55	0.87-2.18-3.61	3.26
AML-RSS	1.69-3.25-5.27	5.64	1.44-2.92-5.06	4.71	1.32-2.74-4.64	4.27	1.31-2.70-4.50	4.20
AML-TOA	2.06-3.74-7.38	6.28	1.52-3.31-5.57	4.93	1.18-2.61-4.27	3.81	1.03-2.38-3.86	3.48
AML-Fusion	1.38-2.91-5.19	4.49	1.15-2.32-3.65	3.49	0.86-1.91-3.39	3.06	0.83-1.83-3.26	2.91
EKF-RSS	0.84-2.22-4.26	3.82	1.06-2.59-4.21	3.81	1.21-2.43-4.07	3.76	1.17-2.55-4.04	3.69
KF-TOA	1.11-2.37-3.95	3.60	1.10-2.06-3.63	3.04	0.81-1.76-2.97	2.53	0.86-1.63-2.95	2.36
EKF-Fusion	0.93-1.90-3.24	2.78	0.86-1.82-3.15	2.59	0.82-1.62-2.62	2.25	0.74-1.49-2.55	2.10
ALPA-RSS	0.82-2.33-4.63	3.88	1.17-2.58-4.30	3.79	1.20-2.48-4.18	3.75	1.21-2.64-4.17	3.78
ALPA-TOA	0.94-2.04-3.33	3.11	0.95-1.90-3.06	2.69	0.72-1.48-2.63	2.52	0.76-1.50-2.64	2.32
ALPA-Fusion	0.84-1.72-2.95	2.58	0.80-1.70-2.85	2.35	0.69-1.37-2.36	2.22	0.70-1.45-2.40	2.08

Table 2. Range estimation error quartiles and RMSE obtained with different algorithms as a function of the number of measurements. All error values are in meters.

Analogously, in Figures 8-9 and Table 3, we summarize the results in position estimation. In this case, we call,[8]

- ML-RSS, ML-TOA, ML-Fusion: the positions obtained with the ML distances and a trilateration technique based on the radical axis of the circles drawn at each anchor's position [10,12-13].
- EKF-RSS, EKF-TOA, EKF-Fusion: the positions obtained by means of an EKF whose measurements model relates the measurements to the target's position.
- PF-RSS, PF-TOA, PF-Fusion: the result of applying the ALPA filter described in Table 1 to the positional-states, with $N=10\ 000$ particles.

8 For the results of Figures 8-9 and Table 3, EKF and ALPA filters use a measurements model that directly relates measurements with positions, avoiding the intermediate step of estimating distances and, therefore, removing the trilateration stage.

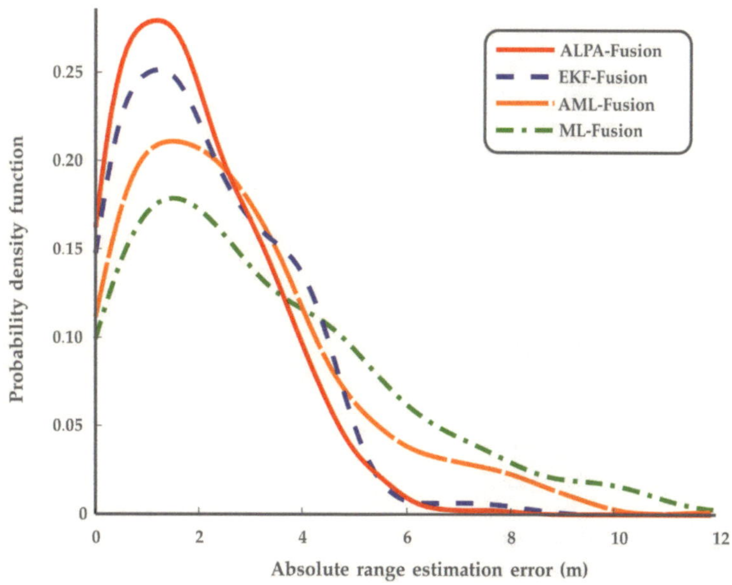

Figure 7. The height and width of the pdf corresponding to the error achieved by the ALPA filter reflect its better performance in comparison to other conventional range estimation techniques. 10 RSS and 10 TOA measurements were taken with respect to each anchor.

	$M=5$		$M=10$		$M=50$		$M=100$	
	Quartiles	**RMSE**	**Quartiles**	**RMSE**	**Quartiles**	**RMSE**	**Quartiles**	**RMSE**
ML-RSS	3.83-5.91-8.49	12.99	3.32-5.21-7.49	8.91	3.35-4.94-6.98	6.64	3.26-5.00-6.60	7.43
ML-TOA	3.95-6.14-8.03	7.64	2.80-4.05-6.64	5.70	2.24-3.34-5.16	4.57	1.63-3.20-4.71	4.09
ML-Fusion	3.15-4.95-7.04	6.73	2.40-3.71-6.11	5.10	1.93-3.03-4.93	4.34	1.64-3.03-4.53	3.89
EKF-RSS	2.94-4.46-6.18	5.11	3.47-4.83-6.85	5.83	3.02-4.12-6.33	5.24	3.02-4.24-6.40	5.24
KF-TOA	1.77-2.79-4.25	3.54	2.05-2.80-3.64	3.11	1.54-2.28-3.09	2.61	1.50-2.22-3.14	2.51
EKF-Fusion	2.20-3.24-4.30	3.50	2.08-2.99-3.90	3.25	1.76-2.32-3.00	2.57	1.71-2.13-2.98	2.41
ALPA-RSS	1.93-3.28-5.18	4.36	3.16-3.91-5.18	4.68	2.38-3.09-4.61	4.23	2.72-3.65-4.97	4.37
ALPA-TOA	1.90-2.59-3.76	3.37	1.63-2.54-3.63	2.98	1.08-1.98-3.25	2.66	1.35-2.18-3.05	2.63
ALPA-Fusion	1.77-2.86-3.46	3.14	1.92-2.61-3.34	2.82	1.23-1.85-3.15	2.49	1.28-2.00-2.64	2.40

Table 3. Position estimation error quartiles and RMSE obtained with several algorithms as a function of the number of measurements. All error values are in meters.

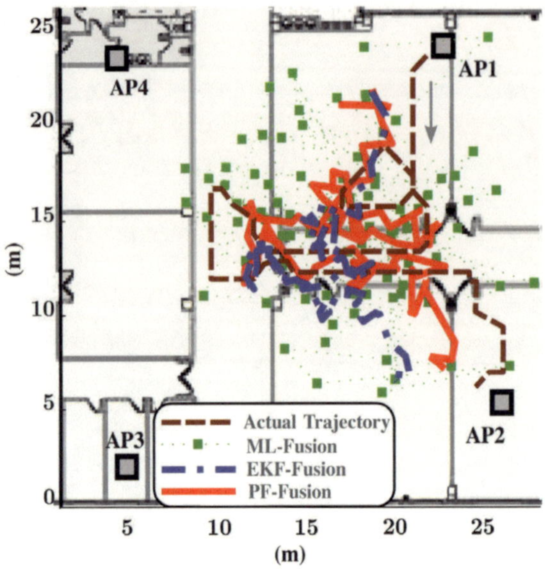

Figure 8. Trajectory followed by the target and position estimates for different positioning methods. 10 RSS and 10 TOA measurements were taken with respect to each anchor.

Figure 9 depicts the pdf of the error in position estimation for the three mentioned RSS/TOA fusion algorithms, taking 10 RSS and 10 TOA measurements in each one of the positions of the target with respect to the four APs. Using only 10 measurements, ML-, EKF- and ALPA-Fusion obtain an error in position estimation lower than 3 meters for 40%, 52%, and 63% of the positions, respectively.

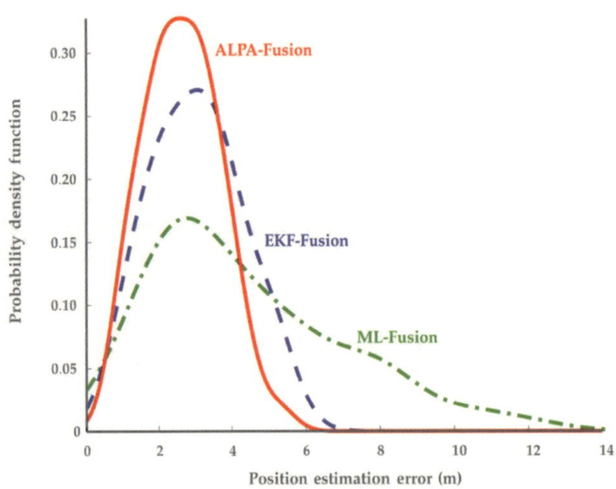

Figure 9. The proposed ALPA filter obtains the best performance with an error lower than 3 meters for more than 63% of the positions.

Figures 8-9 and Table 3 show the better performance of the proposed ALPA filter for all the analyzed scenarios, resulting, for example, in an RMSE of 2.82 meters for the case of only using 10 RSS and 10 TOA measurements, while previous essays obtained RMSEs around 4 meters by using hundreds of measurements [12,28].

5.2. Simulation results

The CRLB provides a lower bound on the minimum achievable mean squared estimation error for any unbiased estimator. In what follows, we use such metric to assess the optimality of the presented ALPA filter against such lower bound.

The Bayesian version of the CRLB is known as the Van Tress CRLB [53], or posterior CRLB, since it is obtained from the posterior distributions of the random state vector [54]. In our case, for each time instant k, the CRLB is,

$$\mathrm{E}\left\{\left(g\left(\mathbf{Z}[k]\right) - \mathbf{y}[k]\right)\left(g(\mathbf{Z}[k]) - \mathbf{y}[k]\right)^T\right\} \succcurlyeq \mathbf{J}_k^{-1} \quad (21)$$

where $g(\mathbf{Z}[k])$ is an unbiased estimator of $\mathbf{y}[k]$ and $\mathbf{J}_k$ is the Fisher information matrix (FIM) obtained as,

$$\mathbf{J}_k = -\mathrm{E}\left\{\nabla_{\mathbf{y}[k]}\left[\nabla_{\mathbf{y}[k]}\log p(\mathbf{Z}[k] \mid \mathbf{y}[k])^{\mathrm{T}}\right]\right\} \quad (22)$$

Tichavský *et al.* proposed a recursive formula to compute the FIM [55]. For the particular case of the linear-Gaussian dynamic model in (3), being $\mathbf{Q}_k$ the covariance matrix in this model, the FIM is given by the recursion [19],

$$\mathbf{J}_{k+1} = \mathbf{J}_{k+1}^{\mathbf{z}} + \left(\mathbf{Q}_k + \mathbf{F}_k \mathbf{J}_k^{-1} \mathbf{F}_k^T\right)^{-1} \quad (23)$$

and

$$\mathbf{J}_{k+1}^{\mathbf{z}} = -\mathrm{E}\left\{\nabla_{\mathbf{y}[k+1]}\left[\nabla_{\mathbf{y}[k+1]}\log p(\mathbf{z}[k+1] \mid \mathbf{y}[k+1])^{\mathrm{T}}\right]\right. \quad (24)$$

To start this recursion, we assume the initial density as Gaussian, then, the initial FIM coincides with its covariance matrix.

Figure 10 compares the RMSE obtained in range estimation by means of the proposed ALPA-Fusion filter with the RMSE obtained by applying the EKF-Fusion method, and with the square root of the CRLB.[9] To obtain such curves, we simulated a trajectory of 85 positions and carried out 1 000 Monte Carlo experiments. Figure 10 again corroborates the remarkable performance of ALPA filter, since the corresponding curve is much closer to the CRLB than the line corresponding to the EKF error.

---

9 We selected a truncated normal distribution as random error to reflect the limited range of the measuring systems. For the proposed adaptive likelihoods, $\mathbf{J}_{k+1}^{\mathbf{z}}$ has no closed-form, then, it was evaluated by Monte Carlo integration.

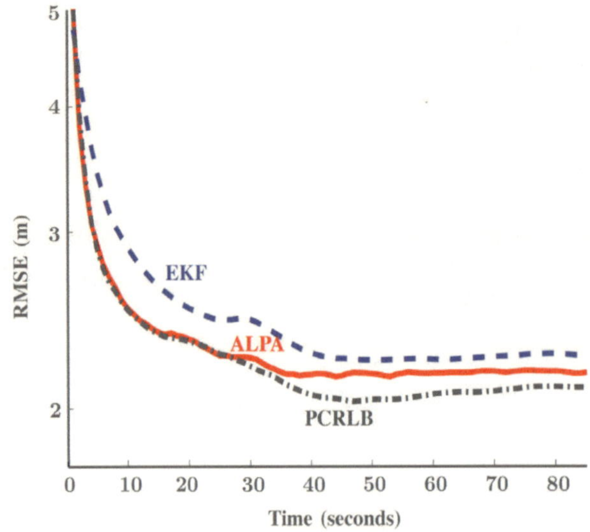

Figure 10. The near-optimal performance of the proposed ALPA filter in harsh environments is corroborated by comparison with the CRLB.

6. Complexity

The key issue in PFs is the exponential growth of computational complexity as a function of the dimension of the state vector, $\mathbf{y}[k]$, whereas EKF grows as the cube of the dimension [56]. For low dimensional problems, PF remains similar to an EKF, however, for high dimensional problems, PFs suffer from the curse of dimensionality [57]. Then, PFs that track ranges instead of positions can be advantageous from a complexity point of view.

Moreover, from Proposition 1, the complexity of the likelihood grows exponentially with the number of samples. However, this complexity can be reduced by removing redundant components from the RSS and TOA pdfs or from the resulting fusion mixture. To this aim, different criteria such as William's criterion [58], Kullback-Leibler distance [59] or clustering [60] can be utilized. Therefore, considering the improvement achieved in range and position estimation with 5 and 10 measurements, the proposed ALPA filter could be a good choice for the designing of VANETs that require low consumption. In these cases, in order to save battery, the OBUs transmit only at discrete intervals; therefore, there is more time available for processing a smaller number of samples.

7. Conclusions

In this chapter we have presented an adaptive likelihood function for robust data fusion in localization systems. Based on this likelihood, we have developed the ALPA filter for range and position estimation. This ALPA filter presents several advantages over conventional techniques,

1. it does not assume any parametric statistical model, utilizing the empirical distribution of the measurements at each time by means of Gaussian kernels;
2. it adaptively fuses RSS and TOA data being extensible to any other type of measurement;
3. it takes advantage of the relationship among positions in time by using Bayesian filtering;
4. it addresses the non-linear and non-Gaussian behavior of the measurements by using particle filtering.

These advantages result in a noticeable improvement with respect to other conventional techniques, as corroborated by the experimental and simulation results. Under NLOS and multipath conditions, ALPA filter obtains not only an RMSE in position estimation lower than 3 meters with only 10 RSS and 10 TOA measurements, but also an error remarkably close to the theoretical benchmark provided by the CRLB.

Therefore, ALPA filter is a valuable choice to provide localization in V2I communication systems. Its extension to cooperative localization would make this localization also possible in VANETs based on V2V communication.

Acknowledgements

This work is partially supported by the Telecommunications Department of the Regional Ministry of Public Works and the Regional Ministry of Education from Castilla y León (Spain), the Spanish projects LEMUR (TIN2009-14114-C04-03) and LORIS (TIN2012-38080-C04-03), and the European Social Fund.

Author details

Javier Prieto[1*], Santiago Mazuelas[2], Alfonso Bahillo[1], Patricia Fernández[1], Rubén M. Lorenzo[1] and Evaristo J. Abril[1]

*Address all correspondence to: javier.prieto@uva.es

1 Dept. of Signal Theory and Communications and Telematics Engineering, University of Valladolid, Valladolid, Spain

2 Laboratory for Information and Decision Systems (LIDS), Massachusetts Institute of Technology, Cambridge, MA, USA

References

[1] Pahlavan, K, Li, X, & Makela, J. Indoor Geolocation Science and Technology. IEEE Communications Magazine (2002). , 40(2), 112-118.

[2] Jarupan, B, & Ekici, E. Location- and Delay-aware Cross-layer Communication in Multihop Vehicular Networks. IEEE Communications Magazine (2009). , 2I

[3] Weimerskirch, A, Hass, J. J, Hu, Y, & Laberteaux, C. K.P. Data Security in Vehicular Communication Networks. In: Hartenstein H., Laberteaux K.P. (ed.) VANET. Vehicular Applications and Inter-Networking Technologies. Chichester: John Wiley & Sons; (2010). , 299-363.

[4] Win, M. Z, Conti, A, Mazuelas, S, Shen, Y, Gifford, M, Dardari, D, & Chiani, M. Network Localization and Navigation via Cooperation. IEEE Communications Magazine (2011). , 49(5), 56-62.

[5] Shen, Y, Wymeersch, H, & Win, M. Z. Fundamental Limits of Wideband Localization-Part II: Cooperative Networks. IEEE Transactions on Information Theory (2010). , 56(10), 4981-5000.

[6] Vegni, A. M, Inzerilli, T, & Cusani, R. Seamless Connectivity Techniques in Vehicular Ad-hoc Networks. In: Almeida M. (ed.) Advances in Vehicular Networking Technologies. Rijeka: In-Tech; (2011). , 3-28.

[7] Gustafsson, F, & Gunnarsson, F. Mobile Positioning Using Wireless Networks. IEEE Signal Processing Magazine (2005). , 22(4), 41-53.

[8] Dardari, D, Conti, A, Ferner, U, Giorgetti, A, & Win, M. Z. Ranging with Ultrawide Bandwidth Signals in Multipath Environments. Proceedings of the IEEE (2009). , 97(2), 404-426.

[9] Falsi, C, Dardari, D, Mucchi, L, & Win, M. Z. Time of Arrival Estimation for UWB Localizers in Realistic Environments. EURASIP Journal on Applied Signal Processing (2006). , 2006-1.

[10] Bahillo, A, Fernández, P, Prieto, J, Mazuelas, S, Lorenzo, R. M, & Abril, E. J. Distance Estimation Based on 802.11 RTS/CTS Mechanism for Indoor Localization. In: Almeida M. (ed.) Advances in Vehicular Networking Technologies. Rijeka: In-Tech; (2011). , 217-236.

[11] Kushki, A, Plataniotis, K. N, & Venetsanopoulos, A. N. Kernel-Based Positioning in Wireless Local Area Networks. IEEE Transactions on Mobile Computing (2007). , 6(6), 689-705.

[12] Mazuelas, S, Bahillo, A, Lorenzo, R. M, Fernández, P, Lago, F. A, García, E, Blas, J, & Abril, E. J. Robust Indoor Positioning Provided by Real-Time RSSI Values in Un-

modified WLAN Networks. IEEE Journal of Selected Topics in Signal Processing (2009). , 3(5), 821-831.

[13] Caffery, J. J. Wireless Location in CDMA Cellular Radio Systems. Norwell: Kluwer Academic Publishers; (1999).

[14] Rappaport, T. S, Reed, J. H, & Woerner, B. D. Position Location Using Wireless Communications on Highways of the Future. IEEE Communications Magazine (1996). , 34(10), 33-41.

[15] Patwari, N. Hero III A.O., Perkins M., Correal N.S., O'Dea J. Relative Location Estimation in Wireless Sensor Networks. IEEE Transactions on Signal Processing (2003). , 51(8), 2137-2148.

[16] Shen, Y, & Win, M. Z. Fundamental Limits of Wideband Localization-Part I: A General Framework. IEEE Transactions on Information Theory (2010). , 56(10), 4956-4980.

[17] Gustafsson, F, Gunnarsson, F, Bergman, N, Forssell, U, Jansson, J, Karlsson, R, & Nordlund, P. J. Particle Filters for Positioning, Navigation, and Tracking. IEEE Transactions on Signal Processing (2002). , 50(2), 425-437.

[18] Bar-shalom, Y. Rong Li X., Kirubarajan T. Estimation with Applications to Tracking and Navigation. New York: John Wiley & Sons; (2001).

[19] Ristic, B, Arulampalam, S, & Gordon, N. Beyond the Kalman Filter, Particle Filters for Tracking Applications. Boston: Artech House; (2004).

[20] Mihaylova, L, Angelova, D, Honary, S, Bull, D. R, Canagarajah, C. N, & Ristic, B. Mobility Tracking in Cellular Networks Using Particle Filtering. IEEE Transactions on Wireless Communications (2007). , 6(10), 3589-3599.

[21] Kushki, A, Plataniotis, K. N, & Venetsanopoulos, A. N. Intelligent Dynamic Radio Tracking in Indoor Wireless Local Area Networks. IEEE Transactions on Mobile Computing (2010). , 9(3), 405-419.

[22] Nicoli, M, Morelli, C, & Rampa, V. A Jump Markov Particle Filter for Localization of Moving Terminals in Multipath Indoor Scenarios. IEEE Transactions on Signal Processing (2008). , 56(8), 3801-3809.

[23] Chen, L, & Wu, L. Mobile Positioning in Mixed LOS/NLOS Conditions under Modified EKF Banks and Data Fusion. IEICE Transactions on Communications (2009). EB(4) 1318-1325., 92.

[24] Song, Y, & Yu, H. A New Hybrid TOA/RSS Location Tracking Algorithm for Wireless Sensor Network. In: Proceedings of the 9th International Conference on Signal Processing, ICSP (2008). October 2008, Beijing, China; 2008., 2008, 26-29.

[25] Chen, B. S, Yang, Y, Liao, F. K, & Liao, J. F. Mobile Location Estimator in a Rough Wireless Environment Using Extended Kalman-Based IMM and Data Fusion. IEEE Transactions on Vehicular Technology (2009). , 58(3), 1157-1169.

[26] Qi, Y. Wireless Geolocation in a non-Line-of-Sight Environment. PhD thesis. Princeton University; (2003).

[27] Mazuelas, S, Lago, F. A, Blas, J, Bahillo, A, Fernández, P, Lorenzo, R. M, & Abril, E. J. Prior NLOS Measurement Correction for Positioning in Cellular Wireless Networks. IEEE Transactions on Vehicular Technology (2009). , 58(5), 2585-2591.

[28] Prieto, J, Bahillo, A, Mazuelas, S, Lorenzo, R. M, Fernández, P, & Abril, E. J. Characterization and Mitigation of Range Estimation Errors for and RTT-Based IEEE 802.11 Indoor Location System. Progress in Electromagnetics Research PIERB (2009). , 15-217.

[29] Chen, P. C. Mobile Position Location Estimation in Cellular Systems. PhD thesis. The State University of New Jersey; (1999).

[30] Wylie, M. P, & Holtzman, J. The non-Line of Sight Problem in Mobile Location Estimation. In: Proceedings of the 1996 5th IEEE International Conference on Universal Personal Communications Record, 29 September- 2 October 1996, Cambridge, MA, USA; (1996).

[31] Roos, T, Myllymäki, P, & Tirri, H. A Statistical Modeling Approach to Location Estimation. IEEE Transactions on Mobile Computing (2002). , 1(1), 59-69.

[32] Hatami, A, & Pahlavan, K. QRPp1-5: Hybrid TOA-RSS Based Localization Using Neural Networks. In: Proceedings of the IEEE Global Telecommunications Conference, GLOBECOM'06, 27 November- 1 December 2006, San Francisco, CA, USA; (2006).

[33] Catovic, A, & Shinoglu, Z. The Cramér-Rao Bounds of Hybrid TOA/RSS and TDOA/RSS Location Estimation Schemes. IEEE Communications Letters (2004). , 8(10), 626-628.

[34] Mcguire, M, Plataniotis, K. N, & Venetsanopoulos, A. N. Data Fusion of Power and Time Measurements for Mobile Terminal Location. IEEE Transactions on Mobile Computing (2005). , 4(2), 142-153.

[35] Cappe, O, Moulines, E, & Ryden, T. Inference in Hidden Markov Models. New York: Springer; (2007).

[36] Rabiner, L, & Juang, B. An Introduction to Hidden Markov Models. IEEE ASSP Magazine (1986). , 3(1), 4-16.

[37] Hashemi, H. The Indoor Radio Propagation Channel. Proceedings of the IEEE (1993). , 81(7), 943-968.

[38] Golden, S. A, & Bateman, S. S. Sensor Measurements for Wi-Fi Location with Emphasis on Time-of-Arrival Ranging, IEEE Transactions on Mobile Computing (2007). , 6(10), 1185-1198.

[39] Prieto, J, Bahillo, A, Mazuelas, S, Fernández, P, Lorenzo, R. M, & Abril, E. J. Self-Calibration of TOA/Distance Relationship for Wireless Localization in Harsh Environ-

ments. In: Proceedings of the IEEE International Conference on Communications, ICC'June 2012, Ottawa, Canada; (2012). , 12, 10-15.

[40] Titterington, D, Smith, A, & Makov, U. Statistical Analysis of Finite Mixture Distributions. Chichester: John Wiley & Sons; (1985).

[41] Mclachlan, G. J, & Krishman, T. The EM Algorithm and Extensions. Hoboken: John Wiley & Sons; (2008).

[42] Neapolitan, R. E. Learning Bayesian Networks. Upper Saddle River: Prentice Hall; (2003).

[43] Prieto, J, Bahillo, A, Mazuelas, S, Lorenzo, R, Blas, J, & Fernández, P. NLOS Mitigation Based on Range Estimation Error Characterization in an RTT-based IEEE 802.11 Indoor Location System. In: Proceedings of the IEEE International Symposium on Intelligent Signal Processing, WISP 2009. August 2009, Budapest, Hungary; (2009). , 26-28.

[44] Jourdan, D, Dardari, D, & Win, M. Z. Position Error Bound for UWB Localization in Dense Cluttered Environments. IEEE Transactions on Aerospace and Electronic Systems (2008). , 44(2), 613-628.

[45] Rosenblatt, M. Remarks on Some Nonparametric Estimates of a Density Function. Annals of Mathematical Statistics (1956). , 27(3), 832-837.

[46] Parzen, E. On Estimation of a Probability Density Function and Mode. Annals of Mathematical Statistics (1962). , 33(3), 1065-1076.

[47] Hwang, J, Lay, S, & Lippman, A. Nonparametric Multivariate Density Estimation: A Comparative Study. IEEE Transactions on Signal Processing (1994). , 42(10), 2795-2810.

[48] Prieto, J, Mazuelas, S, Bahillo, A, Fernández, P, Lorenzo, R. M, & Abril, E. J. Adaptive Data Fusion for Wireless Localization in Harsh Environments. IEEE Transactions on Signal Processing (2012). , 60(4), 1585-1596.

[49] Kushki, A. A Cognitive Radio Tracking System for Indoor Environments. PhD thesis. University of Toronto; (2008).

[50] Scott, D. W. Multivariate Density Estimation. New York: John Wiley & Sons; (1992).

[51] Doucet, A, Godsill, S, & Andrieu, C. On Sequential Monte Carlo Sampling Methods for Bayesian Filtering. Statistics and Computing (2000). , 10(3), 197-208.

[52] Bernardo, M, & Smith, A. F. M. Bayesian Theory. Chichester: John Wiley & Sons; (2000).

[53] Van Trees, H. L. Detection, Estimation, and Modulation Theory: Part I. New York: John Wiley & Sons; (1968).

[54] Bergman, N. Recursive Bayesian Estimation, Navigation and Tracking Applications. PhD thesis. Linköping University, Sweden; (1999).

[55] Tichavský, P, Muravchik, C. H, & Nehorai, A. Posterior Cramér-Rao Bounds for Discrete-Time Nonlinear Filtering. IEEE Transactions on Signal Processing (1998)., 46(5), 1386-1396.

[56] Daum, F. Nonlinear Filters: Beyond the Kalman Filter. IEEE Aerospace and Electronic Systems (2005)., 20(8), 57-69.

[57] Daum, F, & Huan, J. Curse of Dimensionality and Particle Filters. In: Proceedings of IEEE Conference on Aerospace. March 1993, Big Sky, MT, USA; (1993)., 1-8.

[58] Williams, J. L. Gaussian Mixture Reduction for Tracking Multiple Maneuvering Targets in Clutter. PhD thesis. Air Force Institute of Technology; (2003).

[59] Huber, M, & Hanebeck, U. Progressive Gaussian Mixture Reduction. In: Proceedings of the 11th International Conference on Information Fusion, FUSION '08. 30 June- 3 July 2008, Cologne, Germany; (2008).

[60] Schieferdecker, D, & Huber, M. Gaussian Mixture Reduction via Clustering. In: Proceedings of the 12th International Conference on Information Fusion, FUSION'09. July 2009, Seattle, WA, USA; (2009)., 6-9.

Chapter 8

Simulation Platform for Performance Analysis of Cooperative Eigenvalue Spectrum Sensing with a Realistic Receiver Model Under Impulsive Noise

Rausley Adriano Amaral de Souza,
Dayan Adionel Guimarães and
André Antônio dos Anjos

Additional information is available at the end of the chapter

http://dx.doi.org/10.5772/55432

1. Introduction

Currently, the assignment of the electromagnetic spectrum for wireless communication systems follows the so-called fixed allocation policy. In this policy, those who are paying for a given portion of the spectrum obtain the license for exclusively use it, in spite of actually not occupying that portion during all time and in the entire coverage area. This fixed allocation policy, along with the large growth in wireless communications systems and services have led to spectrum congestion and underutilization at the same time [1]. With the advent of the cognitive radio (CR) paradigm [2], cognition-inspired dynamic spectrum access techniques come into action by exploring the underutilized portions of the spectrum in time and space, while causing no or minimum harm in the system that owns the license. In this context, the cognitive radio network is called secondary network, whereas the network licensed for using the spectrum is called primary network.

Among the enormous variety of cognitive tasks that a CR can perform, spectrum sensing is the task of detecting holes (whitespaces) in frequency bands licensed to primary wireless networks, for opportunistic use by the secondary network. Although sensing can be performed by each secondary receiver in a non-cooperative fashion, cooperative spectrum sensing, also known as collaborative spectrum sensing [3], has been considered a solution for problems experienced by CR networks in a non-cooperative sensing situation [4].

In this chapter we first review the main concepts related to the cooperative spectrum sensing, with focus on recent techniques based on the eigenvalues of the received signal covariance matrix, namely: the eigenvalue-based generalized likelihood ratio test (GLRT), the maximum-minimum eigenvalue detection (MMED), the maximum eigenvalue detection (MED), and the energy detection (ED). Then we localize the spectrum sensing subject in the context of vehicular networks. Each cognitive radio is modeled according to a real CR receiver architecture and the sensing environment model considers not only channel fading and thermal noise, but also the presence of impulsive noise (IN). We then describe a MATLAB[1]-based simulation platform developed under the above models, aimed at assessing the performance of several spectrum sensing techniques.

2. Spectrum sensing fundamentals

Figure 1 illustrates a spectrum sensing scenario with a secondary cognitive radio network sharing the spectrum licensed to a primary TV broadcast network, a typical situation covered by the recently issued IEEE 802.22 Standard [5]. In this figure, the coverage areas of the primary transmitters PT1 and PT2 are reaching the cognitive radios CR1 and CR2. Thus, both CRs can check if these primary transmitters are active in a given band (channel) and, if not, the secondary network can opportunistically use that spectrum band. At regular time intervals, the CRs must interrupt their transmissions and verify if the channel is still unoccupied by the primary network. If the channel becomes occupied, the secondary network must stop their transmissions and start the search for another vacant channel.

The decision upon the spectrum occupancy can be independently (non-cooperatively) made by each CR or can be reached by means of cooperation among multiple CRs. In the first case, the sensing performance can be drastically affected by the channel between PTs and CRs. As an example, suppose that a given CR is in a deep fading situation or in a shadowed coverage area, like a topographic depression, inside a building with high radio penetration loss, a tunnel, or behind a large building. This CR can erroneously decide that a channel is vacant and start to use it, causing a harmful interference in the primary network. A similar situation occurs if a CR is out of the reach of the primary network coverage, which is the case of CR3 in Figure 1. It can decide that the sensed channel is vacant, causing strong interference in nearby primary terminals (in T3 in this example). On the other hand, in cooperative spectrum sensing multiple CRs can benefit from the spatial diversity and reduce the effect of the abovementioned problems. Under the control of a secondary base station (BS), a number of CRs monitor a given channel and the decision upon the channel occupancy is made based on the information from all cooperating CRs.

Cooperative spectrum sensing can be classified as centralized and distributed, with the possibility of being relay-assisted [6] in both situations, as illustrated in Figure 2.

1 Matlab is a commercial produt sold by The mathworks, Inc,

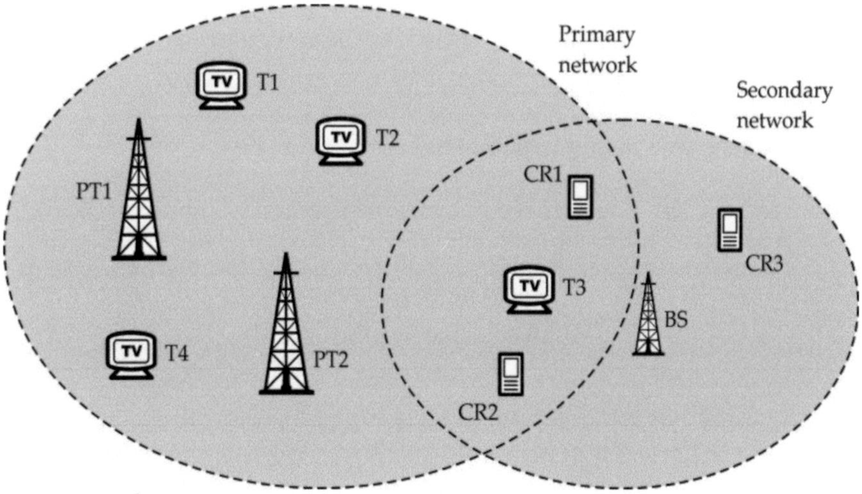

Figure 1. A secondary cognitive radio network opportunistically using the spectrum licensed to a primary TV broadcast network.

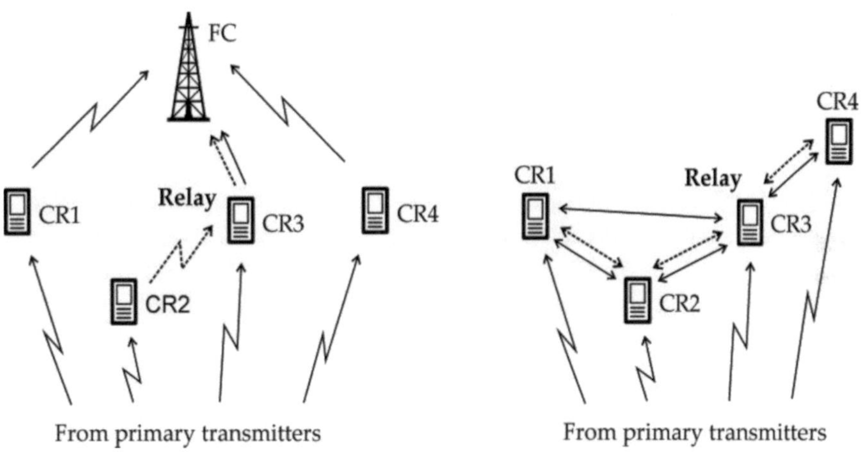

Figure 2. Classification of cooperative spectrum sensing (left) centralized, (right) distributed.

In centralized cooperative sensing, data collected by each cooperating CR (e.g., received samples) are sent via a reporting control channel to a fusion center (FC), in a process called data-fusion. The FC can be a secondary base-station or even a CR. After the FC processes the data from the CRs, it decides upon the occupancy of the sensed channel. Centralized cooperative sensing can also be performed based on the hard decisions made by all cooperating CRs, in a process called decision-fusion. In this case, the binary decisions made by the CRs are combined at the FC using binary arithmetic before the final decision is arrived at. In both centralized schemes, the final decision is reported back to the CRs via a control channel, and an access algorithm takes place in the sequel. In distributed cooperative sensing, no FC exists and the decision is an iterative process performed by the cooperating CRs that communicate among themselves. In the case of relay-assisted cooperative sensing, a given CR may serve as a relay to forward the sensing information from one CR to another, for centralized or distributed operation.

2.1. Spectrum sensing as a binary hypothesis test

The spectrum sensing can be formulated as a binary hypothesis test problem [7], i.e.

$$\begin{aligned} H_0 &: \text{ Primary signal is absent} \\ H_1 &: \text{ Primary signal is present,} \end{aligned} \tag{1}$$

where H_0 is called the null hypothesis, meaning that there is no licensed user signal active in a specific spectrum band, and H_1 is the alternative hypothesis, which indicates that there is an active primary user signal [4].

Two important parameters associated with the assessment of the spectrum sensing performance are the probability of detection, P_{d}, and the probability of false alarm, P_{fa}, which are defined according to

$$\begin{aligned} P_{\mathrm{d}} &= \Pr\{\text{decision} = H_1 \mid H_1\} = \Pr\{T > \gamma \mid H_1\} \\ P_{\mathrm{fa}} &= \Pr\{\text{decision} = H_1 \mid H_0\} = \Pr\{T > \gamma \mid H_0\}, \end{aligned} \tag{2}$$

where $\Pr\{\cdot\}$ is the probability of a given event, T is the detection-dependent test statistic and γ is the decision threshold. The value of γ is chosen depending on the requirements for the spectrum sensing performance, which is typically evaluated through receiver operating characteristic (ROC) curves that show P_{fa} versus P_{d} as they vary with the decision threshold γ.

3. Detection techniques for spectrum sensing

In the context of spectrum sensing, the detection technique aims at extracting from the received signal a test statistic from which the spectrum occupancy is checked, as shown in (2). An overview of some of these techniques is given below:

- *Matched filter detection*: This technique maximizes the signal-to-noise ratio of the received signal and is considered the optimal one if the CR has prior knowledge about primary transmitted signal characteristics, such as the modulation order and type and the pulse shape [8]. If the channel is not a pure additive white Gaussian noise (AWGN) channel, the knowledge of the channel impulse response is needed as well. A matched filter has a challenging practical limitation which is related to the need of estimating or knowing *a priori* the abovementioned information. If such information is not sufficiently accurate, the spectrum sensing performs poorly.

http://dx.doi.org/10.5772/55432

- *Energy detection*: If prior information about the primary transmitted signal is unknown, the energy detection (ED) technique is the optimal one [9], [10]. After the received signal is filtered with a band-pass filter in order to select the desired bandwidth, it is squared and integrated over the sensing interval. The result, which is the test statistic, is compared with a decision threshold so that the absence or presence of the primary signal is inferred. Since this decision threshold depends on the thermal noise variance (noise power), even small noise variance estimation errors can lead to noticeable performance degradation.
- *Cyclostationary detection*: When the primary transmitted signal exhibits cyclostationarity, it can be detected by exploring the periodic behavior of the cyclostationary parameter. This method is more robust to noise uncertainty than energy detection [4],[11],[12]. Although a cyclostationary signal can be detected at lower signal-to-noise ratios compared to other detection strategies, cyclostationary detection is more complex than ED. Moreover, similar to the case of the matched filter detection, it requires some prior knowledge about the primary signal.
- *Eigenvalue-based detection*: Among the existing spectrum sensing detection techniques [6], eigenvalue-based schemes are receiving a lot of attention [13]-[15], mainly because they do not require prior information on the transmitted signal. In some eigenvalue-based schemes, the knowledge of noise variance is not needed either [15]. In eigenvalue spectrum sensing, the test statistic is computed from the eigenvalues of the received signal covariance matrix.

In this chapter we focus in the eigenvalue-based detection. The following techniques are addressed: the eigenvalue-based generalized likelihood ratio test (GLRT), the maximum-minimum eigenvalue detection (MMED), also known as the eigenvalue ratio detection (ERD), the maximum eigenvalue detection (MED), also known as Roy's largest root test (RLRT), and the energy detection (ED) [15]. Although ED is not an exclusively eigenvalue-based detection technique, it can be implemented using eigenvalue information.

Before presenting the specifics of the above eigenvalue-based detection techniques, a discussion about the spectrum sensing in the context of vehicular networks is in order. After this discussion the reader will hopefully be able to analyze the pros and cons of each detection strategy for applications in the vehicular network scenario.

4. Spectrum sensing in the scenario of vehicular networks

4.1. Vehicular Networks

The automotive industry is increasingly incorporating new functionalities to automobiles, aiming at improving the user's experience and safety. Some examples are the use of sensors to detect the proximity of another vehicle or object, automatic headlights and windshield wipers, speed alerts and speed limiters, cruise, traction and braking controls, etc. The majority of these functionalities, till now restricted to control mechanisms and to the interaction between the vehicle and the driver, are based on sensors, actuators and data processing tasks that make some decision or even provide some useful information.

Recently, there have been increased research efforts directed to the development of systems to support the interaction among different vehicles as well. One product of these efforts is the Intelligent Transportation System – ITS [16], which is intended to provide, among other things: traffic information, collision avoidance and congestion control in road transportation systems, as well as interfaces with other transportation systems.

Among the key elements of such intelligent transportation systems, communications play a vital role. A new class of wireless communications networks has emerged in this scenario: the vehicular *ad-hoc* network (VANET). It can be formed with vehicles, or with vehicles and a fixed communication infrastructure nearby the roads or streets. The applications of the VANETs can be focused mainly on *safety*, *entertainment* or *driver assistance* aspects:

- *Safety*: The promise for safety has been one of the main motivators for the development of vehicular networks. In general, the goal is to reduce the number and severity of the accidents by presenting related information to the driver or by actuating on some safety mechanism. Safety applications, however, pose some restrictions to the reliability and latency of the information presentation and actuations. Moreover, these applications must be robust against the presentation of false information and must be able to deal with conflicting data or decisions. Safety applications also must be able to deal with bad or unexpected situations, like the exchange of information about accidents on the road, poor visibility and slippery road, among others.
- *Entertainment*: Most of the applications oriented to entertainment in VANETs are associated to Internet access. This is because users are becoming more dependent on the Internet and they want to access it anywhere and anytime. Among the envisioned entertainment applications are short message services, music and video sharing, and multimedia distribution. In this sort of applications, latency is not a critical issue.
- *Driver assistance*: The main goal of this kind of application is to assist the driver by providing useful information, such as the availability of parking lots, locations in maps, alternative routes and touristic points.

Vehicular networks have some characteristics that distinguish them from other mobile wireless networks. Some of these characteristics are attractive to the network design; others represent important challenges. In the following items we present some of the attractive characteristics [17]; some challenges are listed in the sequel:

- *High transmit power*: Transmit power is not a strong restriction in VANETs, since the vehicle battery or outlets at the road's margins are capable of providing the necessary energy for the mobile and fixed parts of the network, respectively.
- *High computational power*: In principle, processing units with high computational power have no strong limitation of space and power supply in cars or in road units, which facilitates their implementation.
- *Mobility prediction*: Differently from other mobile networks, where it is practically impossible to predict the position of a device, in VANETs this prediction is simplified due to the fixed structure of roads and streets.

http://dx.doi.org/10.5772/55432

- The main challenges to the development of VANETs are [18]:
- *Large scale*: Differently from other *ad-hoc* networks, which normally assume a fixed and moderate size, VANETs can be extended to an entire road and can include a large number of participants.
- *High mobility*: The dynamic nature of the VANETs nodes is complicated when relative vehicle speeds can reach 300 km/h or even more, or when the vehicle density is largely increased, as happens during rush hours or traffic jams.
- *Fragmentation*: The dynamic nature of the VANETs nodes can also bring forth large gaps between vehicles, resulting in clusters of nodes isolated far apart from each other.
- *Topology and network connectivity problems*: Due to the fact that vehicles can constantly change their positions in a short time-frame, nodes are connected during short time intervals and connection loss increases. This represents a great challenge to the development of the network topology and connectivity management strategies.
- *Inadequate spectral bandwidth*: In spite of the fact that a large standardization effort is being put in the VANETs (e.g. in USA the Federal Communications Commission has allocated a 75 MHz band in the 5.9 GHz range for Dedicated Short Range Communications – DSRC [19]), specialists and researchers are claiming for bandwidths adequate to the increasing number and forms of the envisioned applications [20].

4.2. VANET architectures

The VANET architectures define the way in which nodes are organized and how they communicate. Before discussing about these architectures, it is convenient to establish the roles of two key elements of the network: the road site units (RSU) and the mobile nodes or vehicles [21]. The RSUs are part of the network infrastructure, and are typically located nearby roads and streets. They are responsible for providing communication within the VANET and between the VANET and other networks, like the Internet. These units can be public or belong to private service providers. Vehicles have a very important role in VANETs, as they are able to carry information to distant points in the network. Moreover, they are the main observers of the network surrounding environment, since accidents, traffic jams and other traffic events can occur out of reach of the fixed part of the network. In order for the vehicles to be able to monitor the road and traffic conditions, they must be equipped with some key components like sensors, processing and storage units, communication and positioning facilities and an appropriate user interface.

Basically, there are three main vehicular network architectures: pure *ad-hoc* (called vehicle-to-vehicle *ad-hoc* network – V2V), infrastructured[2] (also known as V2I) and hybrid. In V2V, vehicles communicate without any external support or centralizing element. In this architecture, vehicles act as routers or relays, forwarding data through multiple hops. The V2V

2 Althoug the VANET denomination suggests an exclusive association to ad-hoc networks, it has been used in the context of infrastrutured networks as well.

architecture is simpler, since it does not require infrastructure, but it is strongly influenced by vehicle density and mobility pattern, which can cause severe connectivity problems.

To avoid connectivity problems, the V2I architecture adopts static nodes distributed along the roads or streets. These nodes act as access points in IEEE 802.11 networks in structured mode, and centralize the network traffic, also serving as intermediate nodes for communication. The drawback of the V2I is the need of a large number of fixed elements installed along the road or street to improve connectivity, which increases system cost.

The hybrid architecture combines V2V and V2I. In the hybrid mode, a minimum fixed structure is used to increase connectivity e provide interconnection services. Besides the communication among vehicles and fixed parts, the hybrid architecture brings also the possibility of communication between vehicles via single or multiple hops.

4.3. Challenges and opportunities for spectrum sensing in the context of VANETs

One of the biggest challenges of large scale deployment of vehicular networks is the inadequate bandwidth assigned to it [20],[22]. Recent efforts have been put in the assignment of tens of megahertz in the 5.9 GHz band, which is the case of USA, Japan and Europe. However, specialists and researchers are claiming for more bandwidth for emerging services, like [22]:

- *Collision avoidance*: Signals with short durations (large bandwidths) are necessary for improving positioning and ranging, which are fundamental to avoid collisions.
- *Combating mobility*: High moving speeds can give rise to high Doppler spreads in the transmitted signal. This phenomenon can be alleviated with the use of high bandwidth signals, which will be proportionally less affected by a given Doppler spread than narrowband signals.
- *Large volume of data*: Besides information on traffic, road condition, accidents and so on, data from Internet access can increase the demand for high data rates, also increasing the demand for higher bandwidths.

Recent studies have shown that around 70% of the spectrum assigned to TV broadcast in US is underutilized, mainly in small cities and in rural areas [23]. In roads nearby these places the underutilization can be even higher. A solution to the problem of spectrum underutilization and fixed spectrum allocation policy is to incorporate the cognitive radio concept into the context of vehicular networks. To this union it has been given the name Cog-VANET [22]. Similarly to other cognitive radio systems, Cog-VANETs face the challenge of having reliable and fast enough spectrum sensing capabilities. This challenge can be alleviated with the use of cooperative spectrum sensing, which is particularly promising in the Cog-VANET scenario due to the high diversity produced by the high mobility of vehicles [23],[24].

The benefits of the cooperative spectrum sensing to the Cog-VANETs were investigated in [20], [22] and [25]. In [20] the authors propose a CR-based architecture in which sensing information gathered by the vehicles is sent to fixed units at the road side (also called road units). The road units then send the sensing information to a central processing unit. In [25], it is suggested a new frame structure in which the spectrum sensing is considered in a coordinated fashion.

http://dx.doi.org/10.5772/55432

The coordination made by road units determines a group of channels to be sensed by each vehicle in the network. Since in both [20] and [25] it is assumed the presence of several fixed structures alongside the road, the cost can be prohibitive. On the other hand, in [22] the authors suggest a decentralized cooperative sensing in which each vehicle combines different sensing information from other vehicles in order to reach to a decision upon the occupancy of the sensed channel. However, due to the fact that the topology adopted is a V2V, there are inherent connectivity problems, as already mentioned in the previous subsection.

From above one can infer that a possible solution for the problem of spectrum sensing in cognitive vehicular networks would be a mix of centralized and decentralized (or distributed) techniques. Centralized ones would be preferred in regions with fixed structures nearby the roads or streets, whereas decentralized strategies would be preferred in regions without fixed units or with an inadequate density of them. Besides, the hybrid architecture would be the preferred one, since this would permit the communication between vehicles (V2V mode) and between vehicles and fixed units (V2I mode).

The high mobility of vehicles in Cog-VANETs, though advantageous from the perspective of the diversity gain when cooperative sensing is adopted, poses strong requirements in what concerns sensing time, computational power and spectrum agility. Sensing must be fast enough to be effective in this high mobility environment, which in one hand claims for simple detection techniques. On the other hand, the requirement for accuracy claims for more elaborated detection strategies, which in turn can require high computational power in order to keep the sensing time low. Last, radios must be agile in the sense that they must be able to switch between channels quickly. This imposes challenges to the transceiver design. A tradeoff among the above requirements seems to be the best solution, taking into account the centralized or distributed modes of operation that could be changing over time as a given vehicle moves through the network.

5. Centralized cooperative eigenvalue spectrum sensing

In this section we describe the centralized eigenvalue-based cooperative spectrum sensing models, considering the presence of impulsive noise. We first present the conventional model typically adopted in the literature, and then we discuss about a model oriented to the architecture of a real cognitive radio receiver. The impulsive noise models are also addressed in this section. The content of this section is partially based on [26].

Let's consider the well-known baseband memoryless linear discrete-time MIMO fading channel model. Assume that there are m antennas in a CR, or m single-antenna CRs, each one collecting n samples of the received signal from p primary transmitters during the sensing period. Consider that these samples are arranged in a matrix $\mathbf{Y} \in X^{m \times n}$. Similarly, consider that the transmitted signal samples from the p primary transmitters are arranged in a matrix $\mathbf{X} \in X^{p \times n}$. Let $\mathbf{H} \in X^{m \times p}$ be the channel matrix with elements $\{h_{ij}\}$, $i = 1, 2, \ldots, m$ and $j = 1, 2, \ldots, p$, representing the channel gain between the j-th primary transmitter and the i-th sensor (antenna or receiver). Finally, let **V** and $\mathbf{V}_{IN} \in X^{m \times n}$ the matrices containing thermal noise and IN samples that corrupt the received signal, respectively. The matrix of received samples is then

$$\mathbf{Y} = \mathbf{HX} + \mathbf{V} + \mathbf{V}_{IN}. \tag{3}$$

In eigenvalue-based sensing, spectral holes are detected using test statistics computed from the eigenvalues of the sample covariance matrix of the received signal matrix **Y**. If a multi-antenna device is used to decide upon the occupation of a given channel in a non-cooperative fashion, or even in a centralized cooperative scheme with data-fusion, matrix **Y** is formed, and the sample covariance matrix

$$\mathbf{R} = \frac{1}{n}\mathbf{YY}^{\dagger} \tag{4}$$

is estimated, where † means complex conjugate and transpose. The eigenvalues $\{\lambda_1 \geq \lambda_2 \geq \ldots \geq \lambda_m\}$ of **R** are then computed, and assuming a single primary transmitter ($p = 1$), the test statistics for GLRT, MMED, MED, and ED are respectively calculated according to [15]

$$T_{\text{GLRT}} = \frac{\lambda_1}{\frac{1}{m}\text{tr}(\mathbf{R})} = \frac{\lambda_1}{\frac{1}{m}\sum_{i=1}^{m}\lambda_i}, \tag{5}$$

$$T_{\text{MMED}} = \frac{\lambda_1}{\lambda_m}, \tag{6}$$

$$T_{\text{MED}} = \frac{\lambda_1}{\sigma^2}, \tag{7}$$

$$T_{\text{ED}} = \frac{\|\mathbf{Y}\|_F^2}{mn\sigma^2} = \frac{1}{m\sigma^2}\sum_{i=1}^{m}\lambda_i, \tag{8}$$

where σ^2 is the thermal noise power, assumed to be known and with equal value in each sensor input, and tr( ) and $|| \; ||_F$ are the trace and the Frobenius norm of the underlying matrix, respectively.

All the eigenvalue-based detection methods rely on the fact that, asymptotically in n, the sample covariance matrix **R** in the presence of noise only is a diagonal matrix with all its non-zero elements equal to σ^2. Hence, **R** has eigenvalues equal to σ^2 and multiplicity m [27]. In the presence of a primary user, this is no longer true, and these detection methods try to identify this situation: as one can see in (5), in GLRT the ratio between the largest eigenvalue and the average of all the remaining ones is computed; in MMED the ratio between the largest and the smallest eigenvalues is computed; in MED it is assumed that the noise variance σ^2 is known, and the largest eigenvalue is compared with σ^2.

5.1. Conventional model

In the conventional discrete-time memoryless MIMO model (*C-model*), when a centralized cooperative sensing with single-antenna cognitive radios is considered, the matrix with received signal samples **Y** in (3) is presumed to be available at the fusion center as if no signal processing is needed before each row of **Y** is forwarded to the FC by each CR. A simulation setup under *C-model* just considers that **Y** is available to the FC as is.

5.2. Implementation-oriented model

A more realistic model was originally proposed in [28] and called *implementation-oriented model*. It considers typical signal processing tasks performed by each CR before the collected sample values are sent to the FC. The diagram shown in Figure 3 was the main reference for constructing such a model. A wideband band-pass filter (BPF) selects the overall spectrum range to be monitored. The low noise amplifier (LNA) pre-amplifies small signals and a down-conversion (DC) process translates the received signal to in-phase and quadrature (I&Q) baseband signals. The local oscillator (LO) is part of the down-conversion circuitry. A variable-gain amplifier (VGA), which is part of an automatic gain control (AGC) mechanism, is responsible for maintaining the signal within the dynamic range of the analog-to-digital converter (ADC). The channel low-pass filter (LPF) selects the desired spectrum fraction to be sensed. Since filtering affects signal correlation, a whitening process takes place to guarantee that noise samples are decorrelated when the test statistic is computed. This is necessary because the test statistics considered herein implicitly assume decorrelated noise samples.

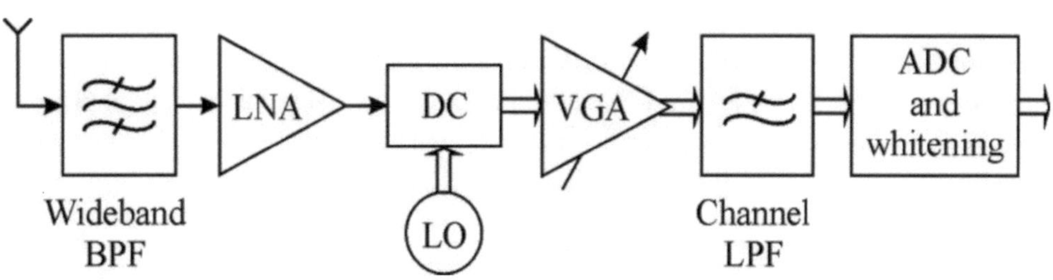

Figure 3. CR receiver diagram (adapted from [28]).

The simulation setup under the realistic implementation-oriented model (*R-model*) has been built to mimic the system diagram shown in Figure 3, in which the down-conversion (DC) to baseband is assumed ideal, as also implicitly assumed in the conventional model. A non-ideal down-conversion in which a direct-conversion receiver (DCR) model is adopted is considered in [28].

Matrices **X**, **H**, **V**, and $\mathbf{V}_{IN}$ under the *R-model* are generated as described in the following subsections.

5.2.1. Transmitted signal

To simulate a Gaussian-distributed noise-like transmitted signal with controllable time correlation, matrix **X** in (3) is formed by filtering independent and identically distributed (i.i.d.) complex Gaussian samples with a length-L moving average (MA) filter with no quantization (using floating-point computation). This type of filter was chosen for reasons of simplicity; any other low-pass filter could be adopted as well. The memory elements in the structure of this and subsequent MA filtering processes are assumed to have zero initial value before the first valid sample is applied to their inputs. As a result, the first $(L - 1)$ samples resulting from the MA filtering, out of $(n + L - 1)$, are discarded before subsequent operations. The Gaussian distribution for the entries of **X** was adopted because it accurately models several modulated signals, for instance the amplitude of a multicarrier signal, such as orthogonal frequency-division multiplexing (OFDM), with a large number of subcarriers, which is the preferred modulation technique in most modern wireless technologies, including several digital television standards. The time correlation introduced by the MA filter models the limited bandwidths of the transmitted and received signals, which are proportional to the symbol rate.

5.2.2. Channel

The elements in the channel matrix **H** in (3) are zero mean i.i.d. complex Gaussian variables that simulate a flat Rayleigh fading channel between each primary transmitter and sensor (cognitive radio), assumed to be constant during a sensing period and independent from one period to another.

5.2.3. Receive filters

To take into account the effect of the CR receive filters on the thermal and impulsive noises, the entries in **V** and $\mathbf{V}_{IN}$ in (3) are MA-filtered complex Gaussian variables that represent, respectively, the colored additive thermal noise and the impulsive noise at the output of the receive filters.

A normalization of filtered samples was done to guarantee the desired received signal-to-noise ratio (SNR), in dB, and the desired average IN power. Specifically, $\mathbf{X} \leftarrow \mathbf{X}/P_X^{1/2}$ for unitary average received signal power, $\mathbf{V} \leftarrow \mathbf{V}\times P_V^{-1/2}\times 10^{-SNR/20}$ for an SNR-dependent average thermal noise power, and $\mathbf{V}_{IN} \leftarrow \mathbf{V}_{IN}\times P_{VIN}^{-1/2}\times K^{1/2}\times 10^{-SNR/20}$ for an average IN power K times the thermal noise power, where $\leftarrow$ represents the normalization process, P_X, P_V, and P_{VIN} are the average time-series powers in **X**, **V**, and $\mathbf{V}_{IN}$ before normalization, respectively. Moreover, to guarantee the desired received SNR, matrix **H** is normalized so that $(1/mp)||\mathbf{H}||_F^2 = (1/mp)\text{tr}(\mathbf{H}^{\dagger}\mathbf{H}) = 1$.

5.2.4. LNA and AGC

The effect of the LNA and the AGC on the samples processed by the i-th CR, $i = 1, 2, \ldots, m$, is given by the gain

$$g_i = \frac{f_{\text{od}} D\sqrt{2}}{6\sqrt{\frac{1}{n}\mathbf{y}_i^{\dagger}\mathbf{y}_i}} = \frac{f_{\text{od}} D\sqrt{2n}}{6\|\mathbf{y}_i\|_2}, \tag{9}$$

where $\mathbf{y}_i$ is the i-th row of $\mathbf{Y}$, i.e., the set of n samples collected by the i-th CR, and $||\mathbf{y}_i||_2$ is the Euclidian norm of $\mathbf{y}_i$. The reasoning behind the definition of these gains is explained as follows: The combined gains of the LNA and the AGC are those that maintain the signal amplitude at the inputs of the in-phase and quadrature ADCs within their dynamic ranges D. By dividing the sample values by the square root of $\mathbf{y}_i^{\dagger}\mathbf{y}_i/n$, which is the average power of $\mathbf{y}_i$, one obtains a sequence with unitary average power. Since $\mathbf{X}$ have complex Gaussian entries, $\{\mathbf{y}_i\}$ have complex Gaussian distributed sample values, conditioned on the corresponding channel gain. If σ^2 is the variance of these samples after the effect of the LNA and the AGC, to guarantee that six standard deviations (practically the whole signal excursion or 99.73% of the sample values) of the I&Q signals will be within $[-D/2, D/2]$, we shall have $6(\sigma^2/2)^{1/2} = D$, which means that the signal power at the output of the AGC will be $\sigma^2 = 2D^2/36$. This justifies the factor $(2^{1/2}D)/6$ in (9). Finally, as the name indicates, the overdrive factor $f_{\text{od}} \geq 1$ is included as a multiplier in (9) to simulate different levels of signal clipping caused by real ADCs, i.e., it produces signal amplitudes greater than or equal to 6σ. For example, an $f_{\text{od}} = 1.2$ means that the dynamic ranges of the signals at the input of the ADCs will be 20% larger than the dynamic ranges of the ADC's inputs. The I&Q clippings act on each sample value s applied to their inputs according to $s \leftarrow \text{sign}(s)\times\min(|s|, D/2)$.

5.2.5. Whitening filter

The Whitening filter matrix $\mathbf{W}$ [29] that multiplies the MA-filtered, amplified and perhaps clipped versions of $\{\mathbf{y}_i\}$ is computed with floating point according to $\mathbf{W} = \mathbf{U}\mathbf{C}^{-1}$, where $\mathbf{U}$ is the orthogonal matrix from $\mathbf{Q} = \mathbf{U}\boldsymbol{\Sigma}\mathbf{K}^{\text{T}}$, the singular-value decomposition of the receive filter covariance matrix $\mathbf{Q}$. The elements of $\mathbf{Q}$ are $Q_{ij} = a_{|i-j|}$, with a_k representing the discrete autocorrelation function of the MA filter impulse response, i.e. $a_k = (1 - k/L)$, for $k \leq L$, and $a_k = 0$ otherwise, for $i, j, k = 0, 1, \ldots, (n-1)$. Matrix $\mathbf{C}$ is the lower triangular matrix from the Cholesky decomposition of $\mathbf{Q}$.

5.2.6. ADC, transmission to the FC and decision

The effect of the analog-to-digital conversion on the processed sample values that will be sent to the FC is modeled by a quantizer with configurable number N_{q} of quantization levels. One must be aware that in practice there will be two ADC operations: the above-described one and the ADC operation with the aim of digital signal processing tasks locally at each CR.

Assuming no bit errors in the reporting channels, the modified received matrix $\mathbf{Y}=\mathbf{HX}+\mathbf{V}+\mathbf{V}_{IN}$ in the implementation-oriented model is then formed at the FC, from which the sample covariance matrix $\mathbf{R}$ is computed, and then the eigenvalues $\{\lambda_i\}, i=1,\ldots,m$. The test statistics for GLRT, MMED, MED, and ED are respectively computed from (5), (6), (7), and (8). In each detection technique, the corresponding test statistic is compared with a threshold computed from the desired false alarm probability, and a final decision upon the occupancy of the sensed channel is reached.

5.3. Impulsive noise model

Impulsive noise can be i) generated from the electrical mains or by direct induction on the receiver, or ii) captured by the receiver antenna. In the first category, the main noise sources are the ignition system of ovens, the control system of dishwasher machines, thermostats of heaters, and switches of fluorescent and incandescent lamps. In the second category, typical sources are lightning and the ignition system of cars, which is particularly relevant in the vehicular network scenario.

Several models are available in the literature for characterizing IN [30]-[34]. Firstly, we discuss about the one presented in [31], in which the IN waveform is generated by properly gating a white noise signal, as illustrated in Figure 4. The main parameters that govern the IN waveform are also shown in this figure. They are configured according to the noise source type, as described in detail in [31].

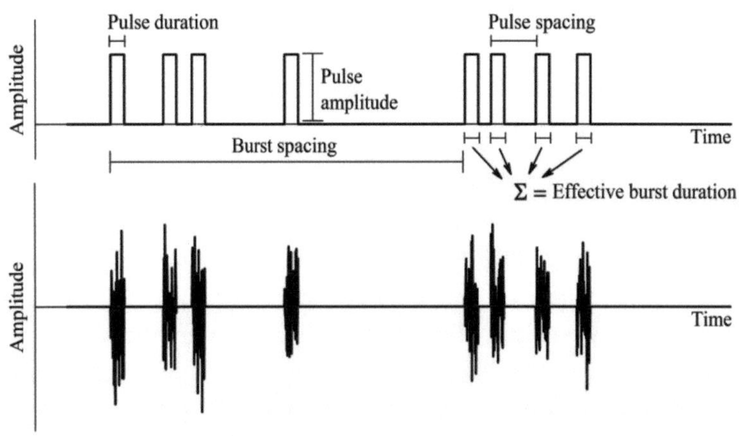

Figure 4. Gating waveform (top) and impulsive noise waveform (bottom) (From [26]).

To adhere the above parameters to the context of spectrum sensing, we have translated them into five other parameters: K is the ratio between the time-series average IN power and the average thermal noise power; p_{IN} denotes the probability of occurrence of IN during a given sensing period, and p_{CR} represents the fraction of CRs hit by IN. As a result, the probability of the occurrence of IN is a Bernoulli random variable with probability of success p_{IN}, and the number of CRs independently hit by IN is a binomial random variable with parameters m and

p_{CR}. A configurable number N_b of IN bursts occurs during a sensing period, each burst having configurable length N_s, i.e., each IN burst corrupts N_s consecutive samples collected by a given CR. The separation between consecutive bursts is uniformly distributed in the discrete-time interval $[0, n - N_b \times N_s]$.

An alternative impulsive noise model was proposed in [33] and [34]. The parameters K, p_{IN} and p_{CR} previously defined are also used in this model, so that it fits to the scenario of cooperative spectrum sensing. The first assumption is the use of an exponential distribution with parameter β in order to settle the interval between impulsive noise pulses. Thus, to compute the number of samples separating each pulse, a random number distributed according to an exponential density with parameter β samples is generated. The pulse amplitudes follow a lognormal distribution. With the purpose of generating the IN amplitude values, a Gaussian random variable Z with mean equal to A dBμV and standard deviation B dB is generated. Then, the value $Z = z$ dBμV is converted to its equivalent amplitude z μV, which is a log-normal variable, using the transformation $z[\mu V] = 10^{z[dB\mu V]/20}$. The phase of the impulsive noise is modeled using a uniformly-distributed random variable θ in $(0, 2\pi]$. Since z is the amplitude, the in-phase and quadrature components of the IN are respectively given by $I_I = z\cos\theta$ and $I_Q = z\sin\theta$.

Figure 5 depicts an example of the compound (thermal plus impulsive) noise waveform (real part) generated according to the model proposed in [33] and [34]. In this figure, the mean of the pulse magnitude $A = 70$ dBμV, the standard deviation of the pulse magnitude $B = 8.5$ dB, $\beta = 900$ samples (which means that it is expected one impulsive noise pulse every 900 samples, on average), Gaussian thermal noise root mean square power 50 dBμV and total generation time of 2 seconds.

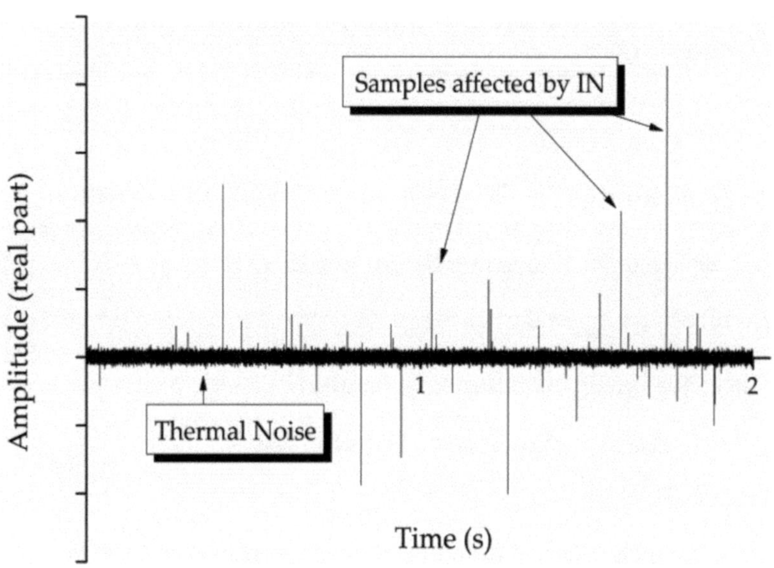

Figure 5. Illustration of the compound (thermal plus impulsive) noise waveform generated by the simulation platform.

6. Description of the simulation platform

This section describes the MATLAB-based simulation platform[3] that provides interactive access to the performance analysis of the four spectrum sensing techniques considered in this chapter: MED, MMED, GLRT and ED. The platform GUI (graphical user interface) is showed in Figure 6.

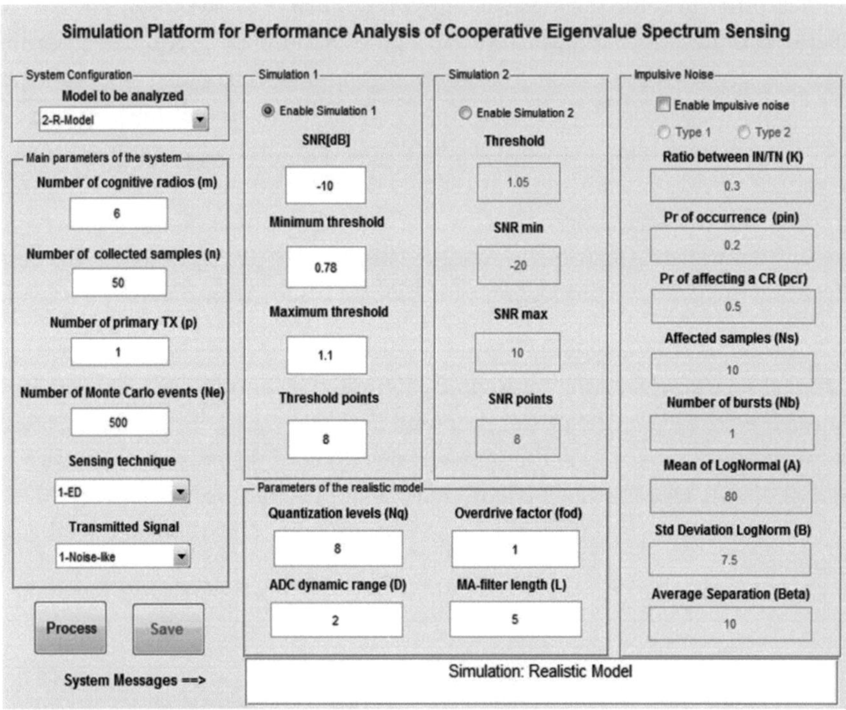

Figure 6. The MATLAB-based simulation platform GUI.

In order to be prompted with the GUI, one must run the file "Main_Spectrum_Sensing.m" in the MATLAB environment. As detailed in the previous sections, there are a large number of scenarios and, consequently, a large number parameters to be set.

First it is necessary to configure the box *System Configuration*, where the model (conventional, *C-model* or the realistic implementation-oriented, *R-model*) can be chosen. In both cases the simulation can be performed with or without impulsive noise.

In the case of the *C-model*, the parameters to be configured are:

- the numbers of CRs (m),
- the number of samples collected from the received signal in each CR (n),

3 Freely downloaded from http://www.inatel.br/lambdaproject. June 2012.

- the number of Monte Carlo simulation events (N_e); the choice of a large N_e will improve the accuracy of the results but will increase the simulation runtime,
- the sensing technique under analysis (MED, MMED, GLRT, ED or user customized sensing technique), and
- the type of transmitted signal (noise, BPSK, QAM or user defined modulation).

If the *R-model* is selected, additional parameters to be configured are:

- the number of quantization levels (N_q),
- the ADC dynamic range (D),
- the overdrive factor (f_{od}), and
- the filter length (L).

If the user intends to simulate the IN influence, the following IN parameters have to be set:

- probability of IN occurrence (p_{IN}),
- fraction of CRs hit by IN (p_{CR}),
- number of samples affected by IN (N_s),
- number of IN bursts (N_b), and
- the ratio between the average IN power and the average thermal noise power (K).

If the user aims to simulate the alternative IN model effect, only the parameter K, p_{IN} and p_{CR} will be used from the previous IN model. Additionally the user has to set:

- the average number of samples between impulsive noise pulses (β),
- the mean of the log-normal impulsive noise amplitudes (A), and
- the standard deviation of the log-normal amplitudes (B).

In the first simulation possibility, called here *Simulation 1*, the user has to configure a fixed SNR value in dB and the decision threshold range (minimum and maximum values). It is also possible to set the number of threshold values (default 8) within the threshold range. This will be the number of points in the graphs prompted at the end of a simulation.

To run the simulation, one must click the *Process* button. There is a countdown timer that appears in the "System Message" in the MATLAB GUI when the simulation is in progress. This counter is used to check the status of the simulation. If it is necessary to stop the processing, just press "*Ctrl* + *C*".

After processing, the simulation platform can return three graphs: P_d and P_{fa} as a function of the threshold range, the receiver operating characteristic (ROC) curve, P_d versus P_{fa} and histograms of the test statistic under the hypothesis H_0 and H_1.

In the second simulation option, named *Simulation 2*, three parameters s have to be defined: a fixed threshold, the minimum and the maximum SNR values and the number of points within the SNR range (default 8). After processing, one graph is prompted, which shows P_d and P_{fa} as a function of the SNR range.

It is possible to save all the generated data by the simulation as a ".dat" file. This file contains three columns: P_d, P_{fa} and the threshold range if the *Simulation 1* is performed; and P_d, P_{fa} and the SNR range if the *Simulation 2* is performed. Additionally, a file with all the system parameter used in the simulations is saved as a ".dat" file. A dialog box enables the user to provide a prefix for the name of both files; they are saved as *"prefix_results.dat"* and *"prefix_parameters.dat"*.

The platform can be easily customized so that the user can choose situations and settings different from the default ones. As an example, are possible to define any type of transmitted signal covariance matrix. Currently, there are three types of such signals: noise, BPSK and MQAM. However, in the *transmitted signal* drop-down list there is a fourth option that can be selected. When this occurs, a other window is prompted to let the user select a ".dat" file, which must contain the desired complex transmit symbols. For example, assume that it is necessary to simulate an MPSK transmitted signal. To do this, simply load a custom ".dat" file with a vector containing the M complex MPSK symbols. In such a way, the user will have flexibility to choose any type of transmitted signal, expanding the capabilities of the simulation platform. Other customization features like this will be included in new versions of the platform, and will be made available at http://www.inatel.br/lambdaproject.

As another example of customization, suppose that it is desired to include a new detection technique besides MMED, MED, GLRT and ED. From the menu *Sensing Technique*, select the option *Customized*. The user will be guided to the module (function) for the generation of the test statistic, called *"Gen_Var_T.m"*, as illustrated in Figure 7. In the box identified as *Customized User Technique* the user must insert the formula for the computation of the new test statistic T, having as inputs any of the variables m, n, $\mathbf{Y}$, $\{g_i\}$, SNR and $\{\lambda_i\}$ defined in Section 5 and commented in the body of the simulation source code.

```
%%%%%% GLRT %%%%%%
case 5
    T = lambda(1) / (sum(diag(W))/m);
%%%%%% Customized User Technique %%%%%%
case 6
    %T = ?;
otherwise
    null; end
```

Figure 7. Part of the simulation source code highlighting the generation of the test statistic *T*.

Different customization features can also be included due to the modular structure of the program exemplified in Figure 7. For example, the sensing performance can be evaluated considering correlated shadowing, fast fading, composite fading, different statistical models of the channel, different types of filters, etc. To perform such additional customization the user must modify the corresponding module of the program. For example, suppose that the user want to assess the spectrum sensing performance in a channel other than the already available flat Rayleigh fading channel. To do that, the user must modify the module *"Gen_Channel_H.m"*. Figure 8 shows the source code for the channel model before (left) and after (right) a customization. For the sake of simplicity and didactical purposes, the new channel model has been chosen as a fixed flat channel with unitary gain. A large number of channel models are available in the literature, but particularly in the case of vehicular networks, the user should refer to [35]-[37] and references therein.

```
function [H] = Gen_Channel_H(m,p)
%Returns Channel array H(mxp) with
%Rayleigh Normalized samples. Input : m
%(number of CR ; p(number of primary
%user) Output: array H (mxn)
%Generates normalized Rayleigh Channel
H = sqrt(2)/2* randn(m,p) +  j *
sqrt(2)/2* randn(m,p);
PH =
1/(m*p)*sum(diag(ctranspose(H)*H));
H = H / sqrt(PH);
```

```
function [H] = Gen_Channel_H(m,p)
%Returns Channel array H(mxp) with
%Rayleigh Normalized samples. Input : m
%(number of CR ; p(number of primary
%user). Output: array H (mxn)
%Generates normalized Rayleigh Channel
%H = sqrt(2)/2* randn(m,p) +  j *
%sqrt(2)/2* randn(m,p);
%PH =
1/(m*p)*sum(diag(ctranspose(H)*H));
%H = H / sqrt(PH);
%%New channel entered by user
H = ones(m,p)
```

Figure 8. Source code of the module responsible for synthesizing the channel model: before customization (left), after a didactical customization (right).

7. Some numeric results about the influence of the system parameters

In this section we present simulation results and a brief discussion concerning the influence of the system parameters under the *C-model* and *R-model* on the spectrum sensing performance for GLRT, MMED, MED, and ED. All the graphs were obtained from the simulation platform described in Section 6.

It is worth mentioning that the ROC curves for all the detection techniques under the *C-model*, for $m = 6$, $n = 50$, and SNR = –10 dB, are in agreement with those reported in [15].

The ROC curves shown hereafter were obtained with a minimum of 15,000 runs in Monte Carlo simulations implemented according to the setup described in Section 5.

Figure 9 shows empirical probability density functions (histograms) of the test statistic generated from the simulation platform using the energy detection and *C-model*, for $m = 6$, $n = 50$, and $p = 1$, and for SNR = −10 dB (left) and SNR = 0 dB (right). The histograms depict the hypotheses H_0 (primary signal absent) and H_1 (primary signal present). One can notice that P_d increases with an increased SNR, considering a fixed threshold, since the area of the histogram on the right of a given threshold under H_1 increases. The net result is an improved (increased) P_d for a given P_{fa} or an improved (reduced) P_{fa} for a given P_d.

Figure 10 shows ROC curves under *C-model* for different values of the number of collected samples (n), relating the probability of false alarm (P_{fa}) and the probability of detection (P_d) for

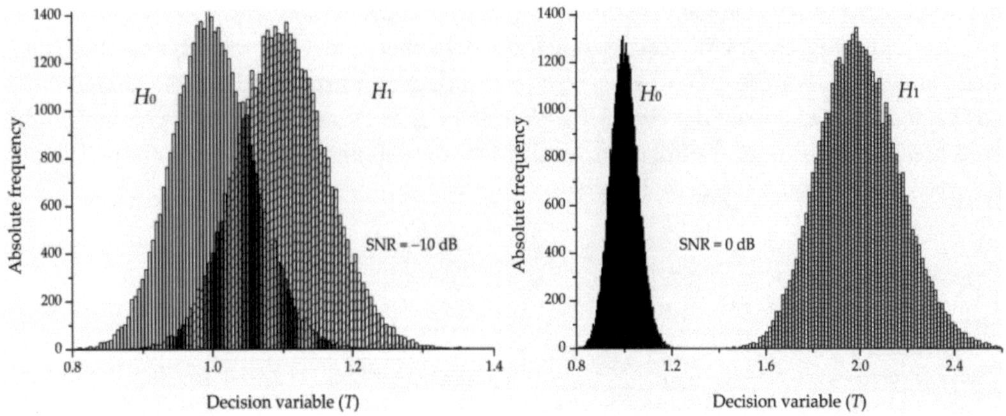

Figure 9. Histograms of the test statistic for ED under different SNR values.

MMED. Clearly, the influence of increasing the number n of collected samples per CR is a performance improvement, considering as fixed the remaining parameters. The threshold range for γ were 2.8 to 4.35 for $n = 60$, 2.8 to 5.1 for $n = 50$, and 2 to 6.5 for $n = 40$. One can notice that the greater the maximum threshold, the smaller P_d and P_{fa}. In the same way, for smaller minimum threshold, P_d and P_{fa} tend to 1. All the fixed remaining parameters are identified in the graph.

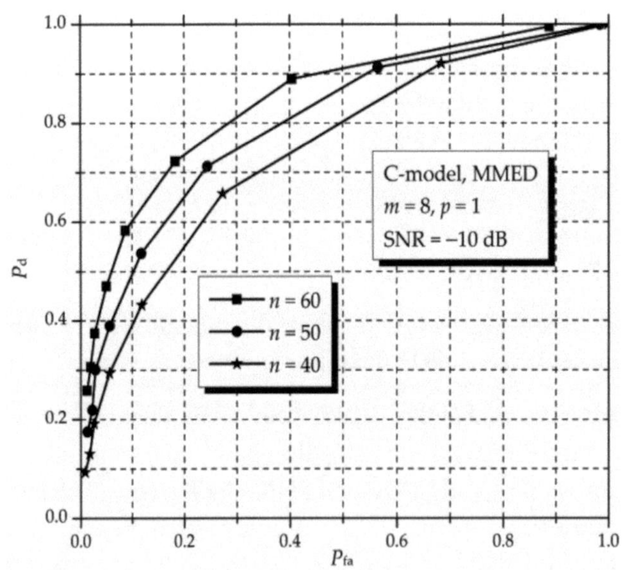

Figure 10. ROC curves for MMED (or ERD) and variable number of collected samples (n) under *C-model*.

In Figure 11 it is plotted two curves under *C-model* for a fixed value of the number of collected samples ($n = 50$), relating the false alarm probability (P_{fa}) and the detection probability (P_d) as a function of the decision threshold γ for MMED. All the remaining parameters are identified in the graph.

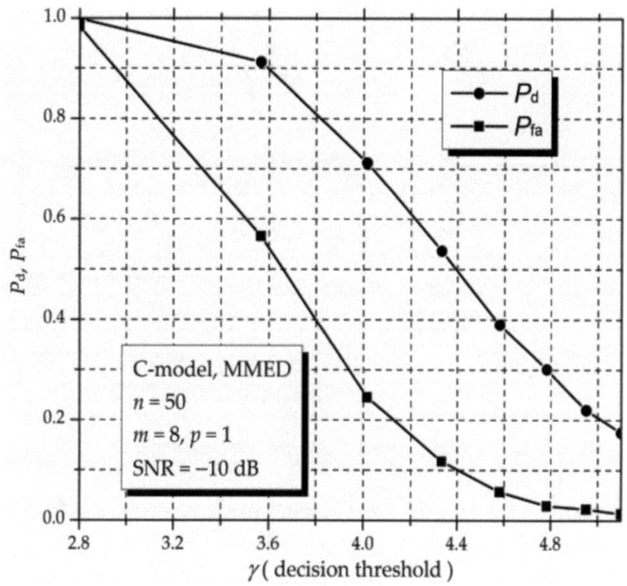

Figure 11. P_d and P_{fa} versus decision threshold γ for MMED (or ERD) for variable number of collected samples (n) under *C-model.*

Now, in order to check the influence of the SNR, Figure 12 depicts some curves under *C-model* for different values of the number of CRs (m), relating the detection probability (P_d) and the signal-to-noise ratio SNR for the RLRT. As expected, the influence of increasing the number of CRs is a performance improvement for a given SNR, considering as fixed the remaining system parameters. On the other hand, one can notice from Figure 12 that for a fixed m, the detection probability increases with the increase of the SNR. All the fixed remaining parameters are identified in the graph.

8. Some numeric results about the influence of the impulsive noise

In this section, further results obtained from the simulation platform are presented and discussed, now for a scenario with IN. We have used only the IN model proposed in [31], since practically the same results were obtained with the alternative model proposed in [33] and [34].

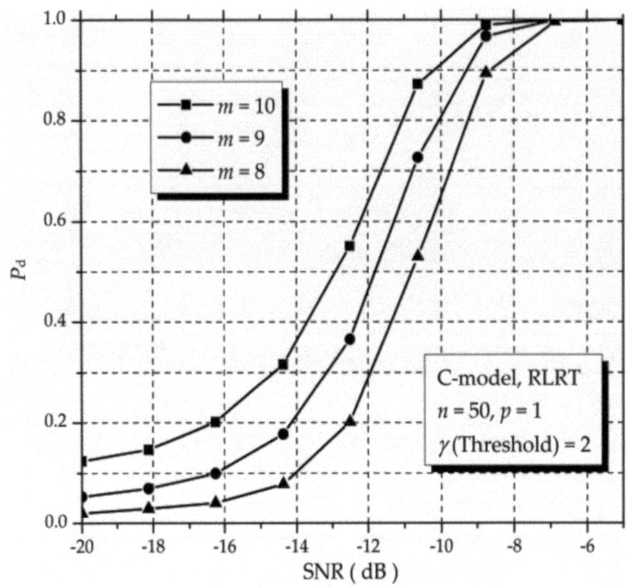

Figure 12. P_d versus SNR for RLRT for variable number of CRs (m) under *C-model*.

As can be seen in Figure 13, under the *C-model* and considering GLRT, impulsive noise degrades the performance for all false alarm probability values. That is, for a fixed P_{fa} the detection probability decreases with an increased impulsive noise power factor K (the ratio between the time-series average IN power and the average thermal noise power). The ranges of decision thresholds used for plotting these ROC curves were 1 to 1.9 for $K = 0$ and $K = 0.5$, and 1 to 2.05 for $K = 1$. All the fixed remaining system parameters are identified in the graph.

In Figure 14, which considers the *C-model* and the GLRT, one can notice that impulsive noise progressively degrades the performance for all false alarm probability values with an increase in the probability of impulsive noise occurrence. The ranges of decision thresholds used for plotting these ROC curves were 1.3 to 1.81 for p_{IN} = 0.2, 1.3 to 1.845 for p_{IN} = 0.4, 1.3 to 1.98 for p_{IN} = 0.6, and 1.3 to 2 for p_{IN} = 0.8. All the remaining fixed parameters are identified in the figure.

Now, in order to show a numerical result with *R-model*, Figure 15 presents some ROC curves considering different values for the number of quantization levels N_q, under IN. The range of decision thresholds used for plotting Figure 15 was 1.28 to 1.82 for all quantization levels N_q. It can be seen that the performance of the sensing scheme is worse for $N_q = 4$, changing slightly from $N_q = 8$ up to $N_q = 32$.

Finally, it is worth mentioning that different values of the system parameters and scenarios can lead to different performances. Thus, a myriad of different scenarios can be exercised through the proposed simulation platform. In this chapter we have just analyzed some parameter variations and scenarios, with the unique objective of showing the applicability of the platform. For a more complete analysis, the reader should refer to [26] and [28].

http://dx.doi.org/10.5772/55432

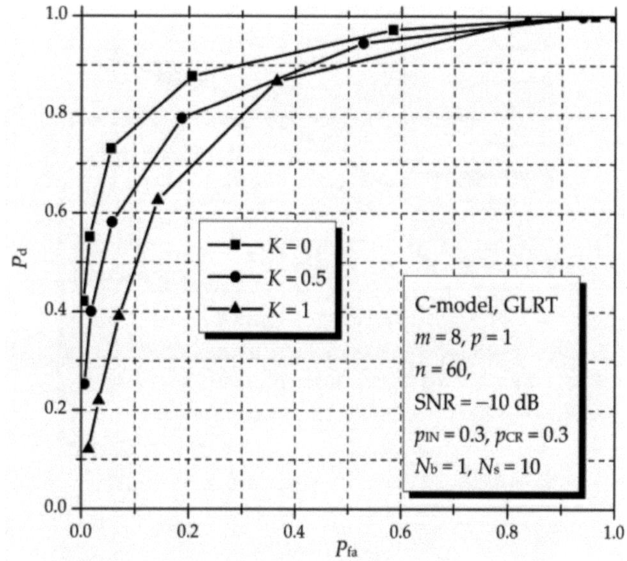

Figure 13. ROC curves for GLRT with and without IN under *C-model*.

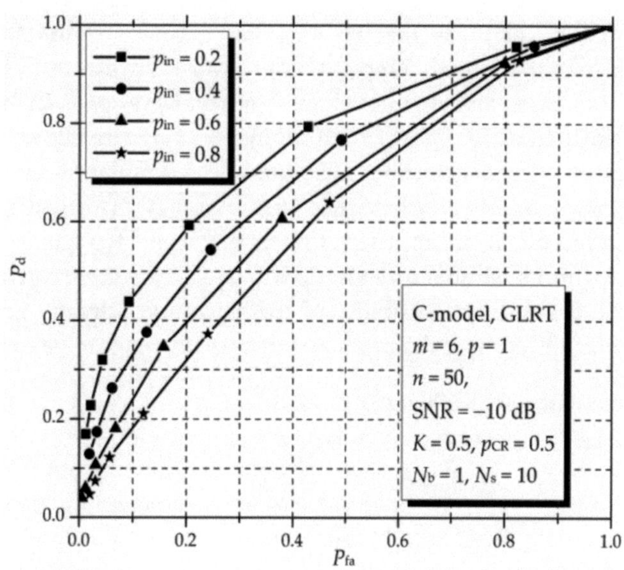

Figure 14. ROC curves for GLRT with IN under p_{IN} variations for C-model.

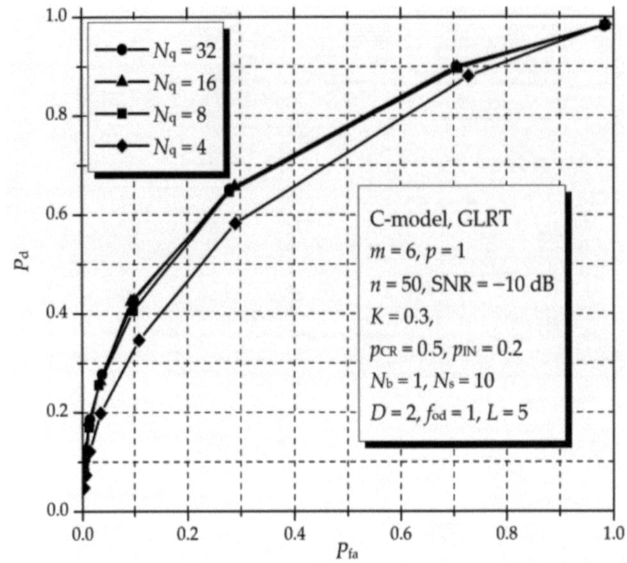

Figure 15. ROC curves for GLRT with IN under N_q variations for *R-model*.

9. Concluding remarks

This chapter described a MATLAB-based simulation platform for assessing the performance of cooperative spectrum sensing in cognitive radio applications such as in vehicular technology applications (e.g. V2V and V2I). The platform is flexible, allowing for the simulation of several sensing techniques, under a broad combination of system parameters. The focus was the cooperative techniques based on the eigenvalues of the received signal, such as GLRT, MMED, and MED, but the customizable feature of the simulation platform allows for the implementation and assessment of new detection techniques. The simulation can be configured to consider a conventional model in which no signal processing is performed by each cooperating CR, or to consider a more realistic approach in which typical CR signal processing tasks are taken into account. It can also simulate a scenario with impulsive noise corrupting the received samples in each CR. Besides the possibility of customizations already implemented in the platform, the modular structure of the program permits new ones to be easily included.

Besides the description of a simulation platform, this chapter reviewed basic concepts related to the spectrum sensing process for cognitive radio applications, and discussed the use of spectrum sensing in the context of cognitive vehicular networks. From this discussion it was possible to infer that the high mobility of vehicles in these networks, though advantageous from the perspective of the diversity gain when cooperative sensing is adopted, poses strong requirements in what concerns sensing time, computational power and spectrum agility. Sensing must be fast enough to be effective in this high mobility environment, which in one hand claims for simple techniques like the energy detection. However, the sensitivity of the

http://dx.doi.org/10.5772/55432

energy detection to noise variance uncertainty can drastically degrade its performance. On the other hand, the requirement for accuracy claims for more elaborated detection strategies that do not need noise variance estimation, as is the case of some eigenvalue-based ones, which in turn can require high computational power to keep the sensing time low. The balance between low complexity and accuracy of the spectrum sensing process is shown to be a formidable research challenge in the scenario of vehicular networks, opening the possibility of new ideas or the improvement of old ones.

Author details

Rausley Adriano Amaral de Souza*, Dayan Adionel Guimarães and
André Antônio dos Anjos

*Address all correspondence to: rausley@inatel.br, dayan@inatel.br, andre-anjos@inatel.br

National Institute of Telecommunications (Inatel), Santa Rita do Sapucaí – MG, Brazil

References

[1] FCC, ET Docket No 03-222 Notice of proposed rule making and order, December 2003.

[2] Mitola J, Maguire GQ. Cognitive radio: making software radios more personal. Personal Communications, IEEE. 1999;6(4):13-8.

[3] Ghasemi A, Sousa ES. Collaborative spectrum sensing for opportunistic access in fading environments. New Frontiers in Dynamic Spectrum Access Networks, 2005. DySPAN 2005. 2005 First IEEE International Symposium on. 2005. p. 131-6.

[4] Letaief KB, Zhang W. Cooperative Communications for Cognitive Radio Networks. Proceedings of the IEEE. 2009;97(5):878-93.

[5] IEEE Standard for Information Technology--Telecommunications and information exchange between systems Wireless Regional Area Networks (WRAN)--Specific requirements Part 22: Cognitive Wireless RAN Medium Access Control (MAC) and Physical Layer (PHY) Specifications: Policies and Procedures for Operation in the TV Bands. 2011.

[6] Akyildiz IF, Lo BF, Balakrishnan R. Cooperative spectrum sensing in cognitive radio networks: A survey. Elsevier Physical Communication. 4(1):40-62.

[7] Yates RD, Goodman D. Probability and Stochastic Processes: A Friendly Introduction for Electrical and Computer Engineers. 2nd ed. 2004.

[8] Sahai A, Hoven N, Tandra R. Some fundamental limits on cognitive radio. Allerton Conference on Control, Communications, and Computation. 2004. p. 1662-71.

[9] Digham FF, Alouini M-S, Simon MK. On the Energy Detection of Unknown Signals Over Fading Channels. Communications, IEEE Transactions on. 2007;55(1):21-4.

[10] Urkowitz H. Energy detection of unknown deterministic signals. Proceedings of the IEEE. 1967;55(4):523- 531.

[11] Cabric D, Mishra SM, Brodersen RW. Implementation issues in spectrum sensing for cognitive radios. Signals, Systems and Computers, 2004. Conference Record of the Thirty-Eighth Asilomar Conference on. 2004. p. 772- 776 Vol.1.

[12] Fehske A, Gaeddert J, Reed JH. A new approach to signal classification using spectral correlation and neural networks. New Frontiers in Dynamic Spectrum Access Networks, 2005. DySPAN 2005. 2005 First IEEE International Symposium on. 2005. p. 144-50.

[13] Zeng Y, Liang Y-C. Eigenvalue-based spectrum sensing algorithms for cognitive radio. Communications, IEEE Transactions on. 2009;57(6):1784-93.

[14] Kortun A, Ratnarajah T, Sellathurai M, Caijun Zhong, Papadias CB. On the Performance of Eigenvalue-Based Cooperative Spectrum Sensing for Cognitive Radio. Selected Topics in Signal Processing, IEEE Journal of. 2011;5(1):49-55.

[15] Nadler B, Penna F, Garello R. Performance of Eigenvalue-Based Signal Detectors with Known and Unknown Noise Level. Communications (ICC), 2011 IEEE International Conference on. 2011. p. 1-5.

[16] ITS Handbook [Internet]. Available from: http://road-network-operations.piarc.org/index.php?option=com_docman&task=cat_view&gid=93&Itemid=39&lang=en

[17] Nekovee M. Sensor networks on the road: the promises and challenges of vehicular ad hoc networks and vehicular grids. Workshop on Ubiquitous Computing and e-Research. Edinburgh, U.K; 2005.

[18] Blum JJ, Eskandarian A, Hoffman LJ. Challenges of intervehicle ad hoc networks. Intelligent Transportation Systems, IEEE Transactions on. 2004;5(4):347- 351.

[19] FCC. News Release, October 1999. Available from http://transition.fcc.gov/Bureaus/Engineering_Technology/News_Releases/1999/nret9006.html.

[20] Fawaz K, Ghandour A, Olleik M, Artail H. Improving reliability of safety applications in vehicle ad hoc networks through the implementation of a cognitive network. Telecommunications (ICT), 2010 IEEE 17th International Conference on. 2010. p. 798-805.

[21] Rawashdeh ZY, Mahmud SM. Communications in Vehicular Ad Hoc Networks, Mobile Ad-Hoc Networks: Applications, Xin Wang (Ed.), ISBN: 978-953-307-416-0, In-

Tech, 2011. Available from: http://www.intechopen.com/books/mobile-ad-hoc-networks-applications/communications-in-vehicular-ad-hoc-networks.

[22] Li H, Irick DK. Collaborative Spectrum Sensing in Cognitive Radio Vehicular Ad Hoc Networks: Belief Propagation on Highway. Vehicular Technology Conference (VTC 2010-Spring), 2010 IEEE 71st. 2010. p. 1-5.

[23] Lennett B. Rural broadband and the TV White space: How unlicensed access to vacante television channels can bring affordable wireless broadband to rural America. New America Foundation: Wireless Future Program, 2008.

[24] Mishra SM, Sahai A, Brodersen RW. Cooperative Sensing among Cognitive Radios. Communications, 2006. ICC '06. IEEE International Conference on. 2006. p. 1658-63.

[25] Wang XY, Ho P-H. A Novel Sensing Coordination Framework for CR-VANETs. Vehicular Technology, IEEE Transactions on. 2010;59(4):1936-48.

[26] Guimarães DA, Souza RAA, Barreto AN. Performance of Cooperative Eigenvalue Spectrum Sensing with a Realistic Receiver Model under Impulsive Noise. Accepted for publication in Journal of Sensor and Actuator Networks. 2012; Dec.

[27] Zeng Y, Liang Y-C. Covariance Based Signal Detections for Cognitive Radio. New Frontiers in Dynamic Spectrum Access Networks, 2007. DySPAN 2007. 2nd IEEE International Symposium on. 2007. p. 202-7.

[28] Guimarães DA, Souza RAA, Implementation-Oriented Model for Centralized Data-Fusion Cooperative Spectrum Sensing, Communications Letters, IEEE. 2012; 16(11), 1804-07. doi: 10.1109/LCOMM.2012.092112.121614.

[29] Cichocki A, Amari S. Adaptive Blind Signal and Image Processing. John Wiley and Sons, Inc.: Chichester, England, 2002.

[30] Mann I, McLaughlin S, Henkel W, Kirkby R, Kessler T. Impulse generation with appropriate amplitude, length, inter-arrival, and spectral characteristics. Selected Areas in Communications, IEEE Journal on. 2002;20(5):901-12.

[31] Lago-Fernández J, Salter J. Modeling Impulsive Interference in DVB-T: Statistical Analysis, Test Waveforms and Receiver Performance. BBC R&D white paper WHP 080, Apr. 2004.

[32] Middleton D. Non-Gaussian noise models in signal processing for telecommunications: new methods an results for class A and class B noise models. Information Theory, IEEE Transactions on. 1999;45(4):1129-49.

[33] Torio P, Sanchez MG. Generating Impulsive Noise [Wireless Corner]. Antennas and Propagation Magazine, IEEE. 2010;52(4):168-73.

[34] Torio P, Sanchez MG, Cuinas I. An algorithm to simulate impulsive noise. Software, Telecommunications and Computer Networks (SoftCOM), 2011 19th International Conference on. 2011. p. 1-4.

[35] Rasheed H, Rajatheva N. Spectrum Sensing for Cognitive Vehicular Networks over Composite Fading. International Journal of Vehicular Technology. 2011:1-9.

[36] Mecklenbrauker CF, Molisch AF, Karedal J, Tufvesson F, Paier A, Bernado L, et al. Vehicular Channel Characterization and Its Implications for Wireless System Design and Performance. Proceedings of the IEEE. 2011; 99(7):1189-212.

[37] Dhar S, Bera R, Giri RB, Anand S, Nath D, Kumar S. An Overview of V2V Communication Channel Modeling. IJCA Proceedings on International Symposium on Devices MEMS, Intelligent Systems & Communication (ISDMISC). 2011;24-34.

Chapter 9

Dual-Hop Amplify-and-Forward Relay Systems with EGC over M2M Fading Channels Under LOS Conditions: Channel Statistics and System Performance Analysis

Talha Batool and Pätzold Matthias

Additional information is available at the end of the chapter

http://dx.doi.org/10.5772/55333

1. Introduction

The recently growing popularity of cooperative diversity [1–3] in wireless networks is due to its ability to mitigate the deleterious fading effects. By utilizing the existing resources of the network, a spatial diversity gain can be achieved without any extra cost resulting from the deployment of a new infrastructure. The fundamental principle of cooperative relaying is that several mobile stations in a network collaborate together to relay the transmit signal from the source mobile station to the destination mobile station. In the simplest mode of operation, the relay nodes just amplify the received signal and forward it towards the destination mobile station. They can also first decode the received signal, encode it again, and then forward it. In both cases, multiple copies of the same signal reach the destination mobile station, which can be combined to achieve a diversity gain by exploiting the virtual antenna array. In addition to the spatial diversity gain, cooperative relaying promises increased capacity, improved connectivity, and a larger coverage range [4–6].

In this paper, a dual-hop amplify-and-forward configuration has been taken into account, where there exist K mobile relays between the source mobile station and the destination mobile station. In addition, the direct link from the source mobile station to the destination mobile station is also present. Such a configuration in turn gives rise to $K+1$ diversity branches. Thus, the previously mentioned spatial diversity gain is achieved by combining the signal received from the $K+1$ diversity branches at the destination mobile station. Among the most important diversity combining techniques [7], maximal ratio combining (MRC) has been proved to be the optimum one [7]. It is widely acknowledged in the literature that a suboptimal and less complex combining technique, referred to as equal gain combining (EGC), performs very close to MRC [7]. Studies regarding the statistical properties of EGC and MRC in non-cooperative networks over Rayleigh, Rice, and Nakagami fading channels are reported in [8–10]. Furthermore, the performance analysis of the said schemes in terms of the bit error and outage probability over Rayleigh, Rice, and Nakagami fading channels can be found in [11–13]. During the last decade, a large number of researchers devoted their efforts to analyzing the performance of cooperative networks. For example, the performance of dual-hop amplify-and-forward relay networks has been extensively investigated for different types of fading channels in [14–26]. A performance analysis in terms of the average bit error probability (BEP) as well as the outage probability of single-relay dual-hop configurations over Rayleigh and generalized-K fading channels is presented in [14] and [16], respectively. A study pertaining to the asymptotic outage behavior of amplify-and-forward dual-hop multi-relay systems with Nakagami fading channels is available in [19], whereas the diversity order is addressed in [23]. The common denominator in the works [15, 17–24] is that they consider MRC at the destination mobile station, where the authors of [24] have also included in their analysis results when EGC is deployed. Performance related issues in multi-relay dual-hop non-regenerative relay systems with EGC over Nakagami-m channels are investigated in [25, 26].

The success story of cooperative relaying in cellular networks has motivated the wireless communications research community to explore their application possibilities in mobile-to-mobile (M2M) communication systems. Relay-based M2M communication systems find their application in intervehicular systems or in other words vehicle-to-vehicle systems. Spreading the information of any kind of emergency situation on roads can be made possible with the use of relay-based M2M systems. The development of relay-based M2M communication systems, however, requires the knowledge of the propagation channel characteristics. It is well known that the multipath propagation channel can efficiently be described with the help of proper statistical models. For example, the Rayleigh distribution is considered to be a suitable distribution to model the fading channel under non-line-of-sight (NLOS) propagation conditions in classical cellular networks [27–29]; a Suzuki process represents a reasonable model for land mobile terrestrial channels [30, 31], and the generalized-K distribution is widely accepted in radar systems [32, 33]. To model fading channels under NLOS propagation conditions in relay-based M2M communication systems, the double Rayleigh distribution is the appropriate choice (see, e.g., [34, 35] and the references therein). Motivated by the applications of the double Rayleigh channel model, a generalized channel model referred to as the $N*$ Nakagami channel model has been proposed in [36]. Furthermore, an extension from the double Rayleigh channel model to the double Rice channel model that is based on the assumption of line-of-sight (LOS) propagation conditions has been proposed in [37]. The authors of [38] have explored the performance of intervehicular cooperative schemes, and they proposed optimum power allocation strategies assuming cascaded Nakagami fading. The performance of several digital modulation schemes over double Nakagami-m channels with MRC diversity has been studied in [39], whereas the BEP analysis of M-ary phase shift keying (PSK) modulated signals over double Rayleigh channels with EGC can be found in [40].

This article focuses on analyzing the statistical properties of EGC over M2M fading channels under LOS propagation conditions as well as the performance of relay-based networks in such channels. As far as the authors are aware, the statistical analysis of EGC over M2M channels assuming LOS propagation conditions has not been carried out yet. In addition,

http://dx.doi.org/10.5772/55333

the performance analysis of multi-relay dual-hop amplify-and-forward cooperative networks in such fading channels is also an open problem that calls for further work. In many practical propagation scenarios, asymmetric fading conditions can be observed in different relay links. Meaning thereby, LOS components can exist in all, none or just in some few transmission links between the source mobile station and the destination mobile station via K mobile relays. Similarly, the LOS component can be present in the direct link from the source mobile station to the destination mobile station. Thus, in order to accommodate the direct link along with the unbalanced relay links, the received signal envelope at the output of the EG combiner is modeled as a sum of a classical Rice process and K double Rice processes. Here, the classical Rice process and double Rice processes are assumed to be statistically independent. Furthermore, it is assumed that K double Rice processes are mutually independent but not necessarily identically distributed. Analytical approximations are derived for the probability density function (PDF), the cumulative distribution function (CDF), the level-crossing rate (LCR), and the average duration of fades (ADF) of the resulting sum process by exploiting the properties of a gamma process[1]. The analysis of these statistical quantities give us a complete picture of the fading channel, since the PDF can well characterize the channel's envelope distribution, and the LCR along with the ADF provide an insight into the fading behavior of the channel. Several performance evaluation measures, such as the statistics of the instantaneous signal-to-noise ratio (SNR) at the output of the equal gain (EG) combiner, amount of fading (AOF), the average BEP, and the outage probability, are thoroughly investigated in this work. It includes also a discussion on the influence of the number of diversity branches $K+1$ as well as the presence of LOS components in the transmission links on the statistics of M2M fading channels with EGC. The approximate analytical results for the PDF, CDF, LCR, ADF, the average BEP, and the outage probability are compared with those of the exact simulation results to validate the correctness of the proposed approach. From the presented results, it can be concluded that the performance of relay-based cooperative systems improves with the presence of LOS components in the relay links. In addition, if the number $K+1$ of diversity branches increases, the better is the system performance.

This article has the following structure. In Section 2, we present the system model for EGC over M2M fading channels under LOS propagation conditions in amplify-and-forward relay networks. Section 3 deals with the derivation and analysis of approximations for the PDF, CDF, LCR, and ADF of the received signal envelope at the output of the EG combiner. In Section 4, analytical approximations for the PDF as well as the moments of the SNR at the output of the EG combiner, the average BEP, and the outage probability are derived and analyzed. Section 5 studies the accuracy of the analytical approximations by simulations and presents a detailed discussion on the obtained results. Finally, the article is concluded in Section 6.

2. EGC over M2M fading channels with LOS components

In this section, we describe the system model for EGC over narrowband M2M fading channels under isotropic scattering conditions with LOS components in a dual-hop cooperative network. In the considered system, we have K mobile relays, which are connected

[1] The material in this paper was presented in part at the International Conference on Communications, ICC 2010, Cape Town, South Africa, May 2010.

in parallel between the source mobile station and the destination mobile station, as illustrated in Fig. 1. It can be seen in this figure that the direct transmission link from the source mobile station to the destination mobile station is also unobstructed.

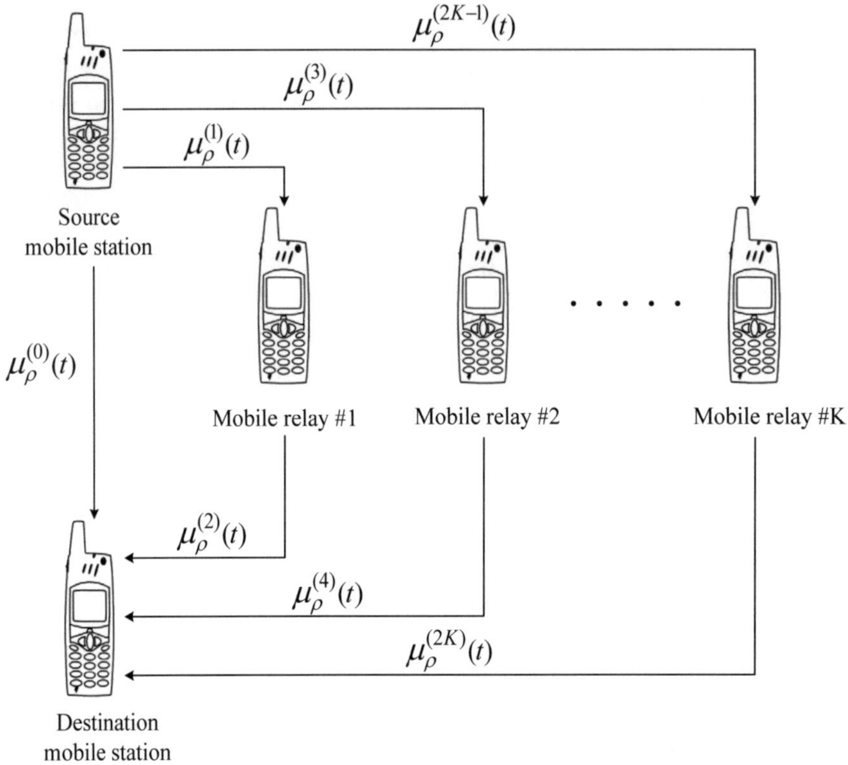

Figure 1. The propagation scenario describing K-parallel dual-hop relay M2M fading channels.

It is assumed that all mobile stations in the network, i.e., the source mobile station, the destination mobile station, and the K mobile relays do not transmit and receive a signal at the same time in the same frequency band. This can be achieved by using the time-division multiple-access (TDMA) based amplify-and-forward relay protocols proposed in [41, 42]. Thus, the signals from the $K + 1$ diversity branches in different time slots can be combined at the destination mobile station using EGC.

Let us denote the signal transmitted by the source mobile station as $s(t)$. Then, the signal $r^{(0)}(t)$ received at the destination mobile station from the direct transmission link between the source mobile station and the destination mobile station can be written as

$$r^{(0)}(t) = \mu_\rho^{(0)}(t)s(t) + n^{(0)}(t) \tag{1}$$

where $\mu_\rho^{(0)}(t)$ models the complex channel gain of the fading channel from the source mobile station to the destination mobile station under LOS propagation conditions. The non-zero-mean complex Gaussian process $\mu_\rho^{(0)}(t)$ comprises the sum of the scattered component $\mu^{(0)}(t)$ and the LOS component $m^{(0)}(t)$ in the direct transmission link from the source mobile station to the destination mobile station, i.e., $\mu_\rho^{(0)}(t) = \mu^{(0)}(t) + m^{(0)}(t)$. In addition, $n^{(0)}(t)$ denotes a zero-mean additive white Gaussian noise (AWGN) process with variance $N_0/2$, where N_0 is the noise power spectral density.

Similarly, we can express the signal $r^{(k)}(t)$ received from the kth diversity branch at the destination mobile station as

$$r^{(k)}(t) = \varsigma_\rho^{(k)}(t)s(t) + n_{\mathrm{T}}^{(k)}(t) \tag{2}$$

where $\varsigma_\rho^{(k)}(t)$ $(k = 1,2,\ldots,K)$ represents the complex channel gain of the kth subchannel from the source mobile station to the destination mobile station via the kth mobile relay under LOS propagation conditions. Furthermore, $n_{\mathrm{T}}^{(k)}(t)\,\forall\, k = 1,2,\ldots,K$ is the total noise in the link from the source mobile station to the destination mobile station via the kth mobile relay. This noise term is analyzed below.

Each fading process $\varsigma_\rho^{(k)}(t)$ in (2) is modeled as a weighted non-zero-mean complex double Gaussian process of the form

$$\varsigma_\rho^{(k)}(t) = \varsigma_{\rho_1}^{(k)}(t) + j\varsigma_{\rho_2}^{(k)}(t) = A_k\,\mu_\rho^{(2k-1)}(t)\mu_\rho^{(2k)}(t) \tag{3}$$

for $k = 1,2,\ldots,K$. In (3), each $\mu_\rho^{(i)}(t)$ is a non-zero-mean complex Gaussian process. For all odd superscripts i, i.e., $i = 2k-1 = 1,3,\ldots,(2K-1)$, the Gaussian process $\mu_\rho^{(i)}(t)$ describes the sum of the scattered component $\mu^{(i)}(t)$ and the LOS component $m^{(i)}(t)$ of the ith subchannel between the source mobile station and the kth mobile relay, i.e., $\mu_\rho^{(i)}(t) = \mu^{(i)}(t) + m^{(i)}(t)$. Whereas, for all even superscripts i, i.e., $i = 2k = 2,4,\ldots,2K$, the Gaussian process $\mu_\rho^{(i)}(t)$ denotes the sum of the scattered component $\mu^{(i)}(t)$ and the LOS component $m^{(i)}(t)$ of the ith subchannel between the kth mobile relay and the destination mobile station. Each scattered component $\mu^{(i)}(t)$ $(i = 0,1,2,\ldots,2K)$ is modeled by a zero-mean complex Gaussian process with variance $2\sigma_i^2$. Furthermore, these Gaussian processes are supposed to be mutually independent, where the spectral properties of each process are characterized by the classical Jakes Doppler power spectral density. The corresponding LOS component $m^{(i)}(t) = \rho_i \exp\{j(2\pi f_\rho^{(i)} t + \theta_\rho^{(i)})\}$ assumes a fixed amplitude ρ_i, a constant Doppler frequency $f_\rho^{(i)}$, and a constant phase $\theta_\rho^{(i)}$.

In (3), A_k is called the relay gain of the kth relay. In order to achieve the optimum performance in a relay-based system, the selection of the relay gain A_k is of critical importance. For fixed-gain relays under NLOS propagation conditions, A_k is usually selected to be [41]

$$A_k = \frac{1}{\sqrt{E\left\{\left|\mu_{\rho\to 0}^{(2k-1)}(t)\right|^2\right\} + N_0}} \tag{4}$$

where $E\{\cdot\}$ is the expectation operator. Notice that $E\left\{\left|\mu_{\rho\to 0}^{(2k-1)}(t)\right|^2\right\} = 2\sigma_{2k-1}^2$ represents the mean power of the NLOS fading channel between the source mobile station and the kth mobile relay. Replacing $\mu_{\rho\to 0}^{(2k-1)}(t)$ by $\mu_{\rho}^{(2k-1)}(t)$ in (4) allows us to express the relay gain A_k associated with LOS propagation scenarios as

$$A_k = \frac{1}{\sqrt{E\left\{\left|\mu_{\rho}^{(2k-1)}(t)\right|^2\right\} + N_0}} = \frac{1}{\sqrt{2\sigma_{2k-1}^2 + \rho_{2k-1}^2 + N_0}}. \tag{5}$$

In practical amplify-and-forward relay systems, the total noise $n_{\mathrm{T}}^{(k)}(t)$ in the link from the source mobile station to the destination mobile station via the kth mobile relay has the following form

$$n_{\mathrm{T}}^{(k)}(t) = A_k\, \mu_{\rho}^{(2k)}(t) n^{(2k-1)}(t) + n^{(2k)}(t) \tag{6}$$

for all $k = 1, 2, \ldots, K$, where $n^{(i)}(t)$ $(i = 1, 2, \ldots, 2K)$ denotes a zero-mean AWGN process with variance $N_0/2$. It is known from the literature (see, e.g., [43] and the references therein) that the total noise $n_{\mathrm{T}}^{(k)}(t)$ can be described under NLOS propagation conditions by a zero-mean complex Gaussian process with variance $N_0 + 2\sigma_{2k}^2 N_0 / \left(2\sigma_{2k-1}^2 + N_0\right)$. It can also be shown that under LOS propagation conditions, the noise process $n_{\mathrm{T}}^{(k)}(t)$ is still a zero-mean complex Gaussian process, but the variance changes to $N_0 + \left(2\sigma_{2k}^2 + \rho_{2k}^2\right) N_0 / \left(2\sigma_{2k-1}^2 + \rho_{2k-1}^2 + N_0\right)$.

Finally, the total signal $r(t)$ at the destination mobile station, obtained after combining the signals $r^{(k)}(t)$ received from $K+1$ diversity branches, can be expressed as

$$r(t) = r^{(0)}(t) + \sum_{k=1}^{K} r^{(k)}(t) = \Xi_{\rho}(t) s(t) + N(t). \tag{7}$$

This result is valid under the assumption of perfect channel state information (CSI) at the destination mobile station. In (7), $\Xi_{\rho}(t)$ represents the fading envelope at the output of the EG combiner, which can be written as [7]

$$\Xi_{\rho}(t) = \xi(t) + \sum_{k=1}^{K} \chi_{\rho}^{(k)}(t) = \left|\mu_{\rho}^{(0)}(t)\right| + \sum_{k=1}^{K} \left|\varsigma_{\rho}^{(k)}(t)\right| \tag{8}$$

where $\xi(t)$ and $\chi_{\rho}^{(k)}(t)$ are the absolute values of $\mu_{\rho}^{(0)}(t)$ and $\varsigma_{\rho}^{(k)}(t)$, respectively. Thus, $\xi(t)$ is the classical Rice process, whereas each of the processes $\chi_{\rho}^{(k)}(t)$ can be identified as a double Rice process. In (7), $N(t)$ is the total received noise, which is given by $N(t) = n^{(0)}(t) + \sum_{k=1}^{K} n_{\mathrm{T}}^{(k)}(t)$.

3. Statistical analysis of EGC over M2M fading channels with LOS components

In this section, we analyze the statistical properties of EGC over M2M fading channels under LOS propagation conditions. The statistical quantities of interest include the PDF, the CDF, the LCR, and the ADF of the stochastic process $\Xi_\rho(t)$ at the output of the EG combiner.

3.1. PDF of a sum of M2M fading processes with LOS components

Under LOS propagation conditions, the received signal envelope $\Xi_\rho(t)$ at the output of the EG combiner is modeled as a sum of a classical Rice process and K independent but not necessarily identical double Rice processes. The PDF $p_{\Xi_\rho}(x)$ of this sum process can be obtained straightforwardly by solving a $(K+1)$-dimensional convolution integral. The computation of this convolution integral is however quite tedious. It can be further shown that the evaluation of the inverse Fourier transform of the characteristic function (CF) does not lead to a simple closed-form expression for the PDF $p_{\Xi_\rho}(x)$. An alternate approach is to approximate $p_\Xi(x)$ either by a simpler expression or by a series. Here, we follow the approximation approach using an orthogonal series expansion. From various options of such series, like, e.g., the Edgeworth series and the Gram-Charlier series, we apply in our analysis the Laguerre series expansion [44]. The Laguerre series provides a good approximation for PDFs that are unimodal (i.e., having a single maximum) with fast decaying tails and positive defined random variables. Furthermore, the Laguerre series is often used if the first term of the series provides a good enough statistical accuracy [44].

The PDF $p_{\Xi_\rho}(x)$ of $\Xi_\rho(t)$ can then be expressed using the Laguerre series expansion as [44]

$$p_{\Xi_\rho}(x) = \sum_{n=0}^{\infty} b_n \, e^{-x} x^{\alpha_L} L_n^{(\alpha_L)}(x) \tag{9}$$

where

$$L_n^{(\alpha_L)}(x) = e^x \frac{x^{(-\alpha_L)} d^n}{x! dx^n} \left[e^{(-x)} x^{n+\alpha_L} \right], \quad \alpha_L > -1 \tag{10}$$

denotes the Laguerre polynomial. The coefficients b_n can be given as

$$b_n = \frac{n!}{\Gamma(n+\alpha_L+1)} \int_0^\infty L_n^{(\alpha_L)}(x) \, p_{\Xi_\rho}(x) dx \tag{11}$$

where $x = y/\beta_L$, and $\Gamma(\cdot)$ is the gamma function [45].

Furthermore, we can obtain the parameters α_L and β_L by solving the system of equations in [44, p. 21] for $b_1 = 0$ and $b_2 = 0$, which yields

$$\alpha_L = \left[\kappa_1^2 / \kappa_2 \right] - 1, \qquad \beta_L = \kappa_2 / \kappa_1 \tag{12a,b}$$

where κ_1 corresponds to the first cumulant (i.e., the mean value) and κ_2 represents the second cumulant (i.e., the variance) of the stochastic process $\Xi_\rho(t)$. Mathematically, we can express κ_n $(n = 1, 2)$ as

$$\kappa_n = \kappa_n^{(0)} + \sum_{k=1}^{K} \kappa_n^{(k)} \tag{13}$$

where $\kappa_n^{(0)}$ are the cumulants associated with the classical Rice process $\xi(t)$. For $n = 1, 2$, the cumulants $\kappa_n^{(0)}$ are equal to [46]

$$\kappa_1^{(0)} = \sigma_0 \sqrt{\frac{\pi}{2}}\, {}_1F_1\!\left(-\frac{1}{2}; 1; -\rho_0^2/2\sigma_0^2\right) \tag{14a}$$

$$\kappa_2^{(0)} = 2\sigma_0^2 \left(1 + \rho_0^2/2\sigma_0^2\right) - \frac{\pi}{2}\sigma_0^2 \left[{}_1F_1\!\left(-\frac{1}{2}; 1; -\rho_0^2/2\sigma_0^2\right)\right]^2 . \tag{14b}$$

In (13), $\kappa_n^{(k)}$ $(k = 1, 2, \ldots, K)$ denote the cumulants of the double Rice process $\chi_\rho^{(k)}(t)$. The mean value and the variance of $\chi_\rho^{(k)}(t)$ are as follows [37]

$$\kappa_1^{(k)} = A_k \sigma_{2k-1}\sigma_{2k} \frac{\pi}{2}\, {}_1F_1\left(-\frac{1}{2}; 1; -\rho_{2k-1}^2/2\sigma_{2k-1}^2\right) {}_1F_1\left(-\frac{1}{2}; 1; -\rho_{2k}^2/2\sigma_{2k}^2\right) \tag{15a}$$

$$\begin{aligned}\kappa_2^{(k)} = &- \left(A_k \sigma_{2k-1}\sigma_{2k} \frac{\pi}{2}\right)^2 \left[{}_1F_1\!\left(-\frac{1}{2}; 1; -\rho_{2k-1}^2/2\sigma_{2k-1}^2\right) {}_1F_1\!\left(-\frac{1}{2}; 1; -\rho_{2k}^2/2\sigma_{2k}^2\right)\right]^2 \\ &+ A_k^2 \left(2\sigma_{2k-1}^2 + \rho_{2k-1}^2\right)\left(2\sigma_{2k}^2 + \rho_{2k}^2\right) .\end{aligned} \tag{15b}$$

It is imperative to stress that here $\kappa_n^{(k)}$ is computed using the expression for A_k given in (5). In (14a) – (15b), ${}_1F_1(\cdot;\cdot;\cdot)$ is the hypergeometric function [45], which can be expanded as

$${}_1F_1\!\left(-\frac{1}{2}; 1; -\frac{\rho_i^2}{2\sigma_i^2}\right) = e^{-\frac{\rho_i^2}{4\sigma_i^2}} \left[\left(1 + \frac{\rho_i^2}{2\sigma_i^2}\right) I_0\left(\frac{\rho_i^2}{4\sigma_i^2}\right) + \frac{\rho_i^2}{2\sigma_i^2} I_1\left(\frac{\rho_i^2}{4\sigma_i^2}\right)\right] \tag{16}$$

where $I_n(\cdot)$ is the nth order modified Bessel function of the first kind [45].

The evaluation of κ_n in (13) is rather straightforward once we have $\kappa_n^{(0)}$ and $\kappa_n^{(k)}$ $(n = 1, 2)$ characterizing $\xi(t)$ and $\chi_\rho^{(k)}(t)$, respectively. Given κ_n, the quantities α_L and β_L can easily be computed using (12a,b). Substituting α_L and β_L in the Laguerre series expansion leads to the exact solution for the PDF $p_{\Xi_\rho}(x)$. Note that the first term of the Laguerre series can be identified as the gamma distribution $p_\Gamma(x)$ [44]. This makes it possible for us to approximate the PDF $p_{\Xi_\rho}(x)$ of $\Xi_\rho(t)$ to the gamma distribution $p_\Gamma(x)$, i.e.,

$$p_{\Xi_\rho}(x) \approx p_\Gamma(x) = \frac{x^{\alpha_L}}{\beta_L^{(\alpha_L+1)}\Gamma(\alpha_L + 1)} e^{-\frac{x}{\beta_L}} . \tag{17}$$

The motivation behind deriving an expression for the PDF $p_{\Xi_\rho}(x)$ of $\Xi_\rho(t)$ is that it can be utilized with ease in the link level performance analysis of dual-hop cooperative networks with EGC. This performance analysis, which results in simple closed-form expressions, is presented in Section 4.

3.2. CDF of a sum of M2M fading processes with LOS components

The probability that $\Xi_\rho(t)$ remains below the threshold level r defines the CDF $F_{\Xi_{\rho-}}(r)$ of $\Xi_\rho(t)$ [47]. After substituting (17) in $F_{\Xi_{\rho-}}(r) = 1 - \int_r^\infty p_{\Xi_\rho}(x)dx$ and solving the integral over x using [45, Eq. (3.381-3)], we can write the CDF $F_{\Xi_{\rho-}}(r)$ in closed form as

$$F_{\Xi_{\rho-}}(r) \approx 1 - \frac{1}{\Gamma(\alpha_L + 1)}\Gamma(\alpha_L, \frac{r}{\beta_L}) \tag{18}$$

where $\Gamma(\cdot,\cdot)$ is the upper incomplete gamma function [45].

3.3. LCR of a sum of M2M fading processes with LOS components

The LCR $N_{\Xi_\rho}(r)$ of $\Xi_\rho(t)$ is a measure to describe the average number of times the stochastic process $\Xi_\rho(t)$ crosses a particular threshold level r from up to down (or from down to up) in a second. The LCR $N_{\Xi_\rho}(r)$ can be computed using the formula [48]

$$N_{\Xi_\rho}(r) = \int_0^\infty \dot{x}\, p_{\Xi_\rho\dot{\Xi}_\rho}(r,\dot{x})d\dot{x} \tag{19}$$

where $p_{\Xi_\rho\dot{\Xi}_\rho}(r,\dot{x})$ is the joint PDF of the stochastic process $\Xi_\rho(t)$ and its corresponding time derivative $\dot{\Xi}_\rho(t)$ at the same time t. Throughout this paper, the overdot represents the time derivative. The task at hand is to find the joint PDF $p_{\Xi_\rho\dot{\Xi}_\rho}(r,\dot{x})$. In Section 3.1, we have shown that the PDF $p_\Xi(x)$ of $\Xi(t)$ can efficiently be approximated by the gamma distribution $p_\Gamma(x)$. Based on this fact, we assume that the joint PDF $p_{\Xi\dot{\Xi}}(r,\dot{x})$ is approximately equal to the joint PDF $p_{\Gamma\dot{\Gamma}}(r,\dot{x})$ of a gamma process and its corresponding time derivative at the same time t, i.e.,

$$p_{\Xi_\rho\dot{\Xi}_\rho}(r,\dot{x}) \approx p_{\Gamma\dot{\Gamma}}(r,\dot{x})\,. \tag{20}$$

A gamma distributed process is equivalent to a squared Nakagami-m distributed process [49]. Thus, applying the concept of transformation of random variables [47, p. 244], we can express the joint PDF $p_{\Gamma\dot{\Gamma}}(x,\dot{x})$ in terms of the joint PDF $p_{N\dot{N}}(y,\dot{y})$ of a Nakagami-m distributed process and its corresponding time derivative at the same time t as

$$p_{\Gamma\dot{\Gamma}}(x,\dot{x}) = \frac{1}{4x}p_{N\dot{N}}\left(\sqrt{x}, \frac{\dot{x}}{2\sqrt{x}}\right)\,. \tag{21}$$

After substituting $p_{N\dot{N}}(y,\dot{y})$ as given in [50, Eq. (13)] in (21), the joint PDF $p_{\Gamma\dot{\Gamma}}(x,\dot{x})$ can be written as

$$p_{\Gamma\dot{\Gamma}}(x,\dot{x}) = \frac{1}{2\sqrt{2\pi x}\dot{\sigma}} \frac{x^{(m-1)}}{(\Omega/m)^m \Gamma(m)} e^{-\frac{x}{(\Omega/m)} - \frac{\dot{x}^2}{8\dot{\sigma}^2 x}} \tag{22}$$

where m, Ω, and $\dot{\sigma}$ are the parameters associated with the Nakagami-m distribution. The result in (22) can be expressed in terms of the parameters of the gamma distribution, i.e., α_L and β_L, as

$$p_{\Gamma\dot{\Gamma}}(x,\dot{x}) = \frac{1}{2\sqrt{2\pi\beta x}} \frac{x^{\alpha_L}}{\beta_L^{(\alpha_L+1)} \Gamma(\alpha_L+1)} e^{-\frac{x}{\beta_L} - \frac{\dot{x}^2}{8\beta x}} \tag{23}$$

For a classical Rayleigh fading channel, β is the negative curvature of the autocorrelation of the inphase and quadrature components of the underlying gaussian processes [46]. Keeping the expression of β for classical Rayleigh channels in mind, we can intuitively equate β with $\pi^2(\kappa_n^{(0)}/\kappa_n^{(0)})(f_{S_{\max}}^2 + f_{D_{\max}}^2) + \pi^2(\sum_{k=1}^K \kappa_2^{(k)} / \sum_{k=1}^K \kappa_1^{(k)})(f_{S_{\max}}^2 + 2f_{R_{\max}}^2 + f_{D_{\max}}^2)$ here. The quantities $f_{S_{\max}}$ and $f_{D_{\max}}$ represent the maximum Doppler frequencies caused by the motion of the source mobile station and the destination mobile station, respectively. For simplicity reasons, the maximum Doppler frequencies caused by the motion of mobile relays are assumed to be equal such that $f_{R_{\max}}^{(1)} = f_{R_{\max}}^{(2)} = \cdots = f_{R_{\max}}^{(K)} = f_{R_{\max}}$.

Finally, substituting $p_{\Xi_\rho\dot{\Xi}_\rho}(r,\dot{x})$ in (19) and solving the integral over $\dot{x}$ using [45, Eq. (3.326-2)], we reach a closed-form solution for the LCR $N_{\Xi_\rho}(r)$, i.e.,

$$N_{\Xi_\rho}(r) \approx \int_0^\infty \dot{x}\, p_{\Gamma\dot{\Gamma}}(r,\dot{x})d\dot{x} = \sqrt{\frac{2r\beta}{\pi}} \frac{r^{\alpha_L} e^{-\frac{r}{\beta_L}}}{\beta_L^{(\alpha_L+1)}\Gamma(\alpha_L+1)} = \sqrt{\frac{2r\beta}{\pi}} p_{\Xi_\rho}(r) \tag{24}$$

which shows that the LCR $N_{\Xi_\rho}(r)$ is approximately proportional to the PDF $p_{\Xi_\rho}(r)$ of $\Xi_\rho(t)$.

3.4. ADF of a sum of M2M fading processes with LOS components

The ADF $T_{\Xi_{\rho-}}(r)$ of $\Xi_\rho(t)$ is the expected value of the time intervals over which the stochastic process $\Xi_\rho(t)$ remains below a certain threshold level r. Mathematically, the ADF $T_{\Xi_{\rho-}}(r)$ is defined as the ratio of the CDF $F_{\Xi_{\rho-}}(r)$ and the LCR $N_{\Xi_\rho}(r)$ of $\Xi_\rho(t)$ [7], i.e.,

$$T_{\Xi_{\rho-}}(r) = \frac{F_{\Xi_{\rho-}}(r)}{N_{\Xi_\rho}(r)}. \tag{25}$$

By substituting (18) and (24) in (25), we can easily obtain an approximate solution for the ADF $T_{\Xi_{\rho-}}(r)$.

The significance of studying the LCR $N_{\Xi_\rho}(r)$ and the ADF $T_{\Xi_{\rho-}}(r)$ of $\Xi_\rho(t)$ lies in the fact that they provide an insight into the rate of fading of the stochastic process $\Xi_\rho(t)$. The knowledge about the rate of fading is essential for both the design as well as the optimization of coding and interleaving schemes to combat M2M fading in the relay links in cooperative networks.

4. Performance analysis in M2M fading channels with LOS components and EGC

This section is dedicated to the system's performance analysis in M2M fading channels with EGC under LOS propagation conditions. The performance evaluation measures of interest include the PDF as well as the moments of the SNR, AOF, the average BEP, and the outage probability.

4.1. Analysis of the SNR

4.1.1. Derivation of the instantaneous SNR expression

We computed the total received signal envelope at the output of the EG combiner $\Xi_\rho(t)$ and the total received noise $N(t)$ in Section 2. Using these results, we can now express the instantaneous SNR per bit $\gamma_{\text{EGC}}(t)$ at the output of the EG combiner as [51, 52]

$$\gamma_{\text{EGC}}(t) = \frac{\Xi_\rho^2(t)}{E\{N^2(t)\}} E_b \tag{26}$$

where E_b is the energy (in joules) per bit and $E\{N^2(t)\}$ is the variance of the noise term at the output of the matched filter. Evaluating $E\{N^2(t)\}$ leads us to

$$\begin{aligned} E\{N^2(t)\} &= E\left\{\left(n^{(0)}(t) + \sum_{k=1}^{K} n_{\text{T}}^{(k)}(t)\right)^2\right\} \\ &= (K+1)N_0 + \sum_{k=1}^{K} \frac{2\sigma_{2k}^2 + \rho_{2k}^2}{2\sigma_{2k-1}^2 + \rho_{2k-1}^2 + N_0} N_0 \,. \end{aligned} \tag{27}$$

4.1.2. PDF of the SNR

The PDF $p_{\gamma_{\text{EGC}}}(z)$ of $\gamma_{\text{EGC}}(t)$ can be obtained using the relation

$$p_{\gamma_{\text{EGC}}}(z) = \frac{1}{(E_b/E\{N^2(t)\})} p_{\Xi_\rho^2}\left(\frac{z}{E_b/E\{N^2(t)\}}\right) \tag{28}$$

where $p_{\Xi_\rho^2}(z)$ is the squared received signal envelope $\Xi_\rho^2(t)$ at the output of the EG combiner, which can be obtained by a simple transformation of the random variables [47, p. 244] as follows

$$\begin{aligned} p_{\Xi_\rho^2}(z) &= \frac{1}{2\sqrt{z}} p_{\Xi_\rho}(\sqrt{z}) \\ &\approx \frac{1}{2\beta_L^{(\alpha_L+1)}\Gamma(\alpha_L+1)} z^{\left(\frac{\alpha_L-1}{2}\right)} e^{-\frac{\sqrt{z}}{\beta_L}}, \quad z \geq 0 \,. \end{aligned} \tag{29}$$

The substitution of (29) in (28) leads us to the approximation for the PDF $p_{\gamma_{\text{EGC}}}(z)$ of $\gamma_{\text{EGC}}(t)$ in the following closed form expression

$$p_{\gamma_{\text{EGC}}}(z) \approx \frac{1}{2(E_b/E\{N^2(t)\})^{\left(\frac{\alpha_L+1}{2}\right)}} \frac{z^{\left(\frac{\alpha_L-1}{2}\right)}}{\beta_L^{(\alpha_L+1)}\Gamma(\alpha_L+1)} e^{-\frac{\sqrt{z}}{\beta_L\sqrt{E_b/E\{N^2(t)\}}}} . \quad (30)$$

4.1.3. Moments of the SNR

Substituting (30) in $m^{(n)}_{\gamma_{\text{EGC}}} = \int_{-\infty}^{\infty} z^n \, p_{\gamma_{\text{EGC}}}(z)dz$ and solving the integral over z using [45, Eq. (3.478-1)] allows us to express approximately the nth moment of the SNR $\gamma_{\text{EGC}}(t)$ in closed form as

$$m^{(n)}_{\gamma_{\text{EGC}}} \approx \beta_L^{2n} \left(\frac{E_b}{E\{N^2(t)\}}\right)^n \frac{\Gamma(\alpha_L+2n+1)}{\Gamma(\alpha_L+1)} . \quad (31)$$

4.1.4. Amount of Fading

The AOF is defined as the ratio of the variance $\sigma^2_{\gamma_{\text{EGC}}}$ and the squared mean value $m^{(1)}_{\gamma_{\text{EGC}}}$ of the SNR $\gamma_{\text{EGC}}(t)$, i.e., [26, 53]

$$\text{AOF} = \frac{\sigma^2_{\gamma_{\text{EGC}}}}{\left(m^{(1)}_{\gamma_{\text{EGC}}}\right)^2} = \frac{m^{(2)}_{\gamma_{\text{EGC}}} - \left(m^{(1)}_{\gamma_{\text{EGC}}}\right)^2}{\left(m^{(1)}_{\gamma_{\text{EGC}}}\right)^2} . \quad (32)$$

Computing the first two moments of $\gamma_{\text{EGC}}(t)$ using (31) and substituting the results in (32) yields the following closed-form approximation for the AOF

$$\text{AOF} \approx \left(\alpha_L^2 + 7\alpha_L + 12\right) \beta_L^2 \Gamma(\alpha_L+1) - 1 . \quad (33)$$

4.2. Average BEP

By way of example, we focus on the average BEP P_b of M-ary PSK modulation schemes. The average BEP P_b over the fading channel statistics at the output of the EG combiner can be obtained using the formula [51]

$$P_b = \int_0^{\infty} p_{\Xi_\rho}(x) P_{b|\Xi_\rho}(x)dx \quad (34)$$

where $P_{b|\Xi_\rho}(x)$ is the BEP of M-ary PSK modulation schemes conditioned on the fading amplitudes $\{x_k\}_{k=0}^K$, and $x = \sum_{k=0}^K x_k$. Here, the fading amplitude x_0 follows the classical

Rice distribution. Furthermore, the fading amplitudes $\{x_k\}_{k=1}^{K}$ are characterized by the double Rice distribution.

The conditional BEP $P_{b|\Xi_p}(x)$ of M-ary PSK modulation schemes can be approximated as [54]

$$P_{b|\Xi_p}(x) \approx \frac{a}{\log_2 M} Q\left(\sqrt{2g\log_2 M\gamma_{\text{EGC}}(x)}\right) \approx \frac{a}{\log_2 M} Q\left(\sqrt{\frac{2g\log_2 M E_b}{E\{N^2(t)\}}x^2}\right) \tag{35}$$

where $M = 2^b$ with b as the number of bits per symbol, and $Q(\cdot)$ is the error function [45]. The parameter a equals 1 or 2 for M-ary PSK modulation schemes when $M = 2$ or $M > 2$, respectively, whereas for all M-ary PSK modulation schemes $g = \sin^2(\pi/M)$ [39].

Substituting (17) and (35) in (34) leads to the approximate solution for the average BEP P_b in the form

$$P_b \approx \frac{a}{\log_2 M} \frac{1}{\beta_L^{(\alpha_L+1)}\Gamma(\alpha_L+1)} \int_0^\infty x^{\alpha_L} e^{-\frac{x}{\beta_L}} Q\left(\sqrt{\frac{2g\log_2 M E_b}{E\{N^2(t)\}}x^2}\right) dx\,. \tag{36}$$

4.3. Outage Probability

The outage probability $P_{\text{out}}(\gamma_{th})$ is defined as the probability that the SNR $\gamma_{\text{EGC}}(t)$ at the output of the EG combiner falls below a certain threshold level γ_{th}. Substituting (30) in $P_{\text{out}}(\gamma_{th}) = Pr\{\gamma_{\text{EGC}} \leq \gamma_{th}\} = 1 - \int_{\gamma_{th}}^{\infty} p_{\gamma_{\text{EGC}}}(z)dz$

$$P_{\text{out}}(\gamma_{th}) \approx 1 - \frac{1}{\Gamma(\alpha_L+1)}\Gamma\left(\alpha_L+1, \frac{\sqrt{\gamma_{th}}}{\beta_L\sqrt{E_b/E\{N^2(t)\}}}\right). \tag{37}$$

5. Numerical results

The aim of this section is to evaluate and to illustrate the derived theoretical approximations given in (17), (24), (25), (36), and (37) as well as to investigate their accuracy. The correctness of the approximated analytical results is confirmed by evaluating the statistics of the waveforms generated by utilizing the sum-of-sinusoids (SOS) method [46]. These simulation results correspond to the true (exact) results here. The waveforms $\tilde{\mu}^{(i)}(t)$ obtained from the designed SOS-based channel simulator are considered as an appropriate model for the uncorrelated Gaussian noise processes $\mu^{(i)}(t)$ making up the received signal envelope at the output of the EG combiner. The model parameters of the channel simulator have been computed by using the generalized method of exact Doppler spread (GMEDS$_1$) [55]. Each waveform $\tilde{\mu}^{(i)}(t)$ was generated with $\text{N}_l^{(i)} = 14$ for $i = 0,1,2,\ldots,2K$ and $l = 1,2$, where

$N_l^{(i)}$ is the number of sinusoids chosen to simulate the inphase ($l = 1$) and quadrature ($l = 2$) components of $\tilde{\mu}^{(i)}(t)$. It is widely acknowledged that the distribution of the absolute value $\left|\tilde{\mu}^{(i)}(t)\right|$ of the simulated waveforms closely approximates the Rayleigh distribution if $N_l^{(i)} \geq 7 \, (l = 1,2)$ [46]. Thus, by selecting $N_l^{(i)} = 14$, we ensure that the waveforms $\tilde{\mu}^{(i)}(t)$ have the required Gaussian distribution. The variance of the inphase and quadrature component of $\mu^{(i)}(t)$ ($\tilde{\mu}^{(i)}(t)$) is equal to $\sigma_i^2 = 1 \, \forall \, i = 0,1,2,\ldots,2K$, unless stated otherwise. The maximum Doppler frequencies caused by the motion of the source mobile station, K mobile relays, and the destination mobile station, denoted by $f_{\mathrm{S_{max}}}$, $f_{\mathrm{R_{max}}}$, and $f_{\mathrm{D_{max}}}$, were set to 91 Hz, 125 Hz, and 110 Hz, respectively. The total number of symbols generated for a reliable evaluation of the BEP curves was 10^7.

In this section, we have attempted to highlight the influence of a LOS component on the statistics of the received signal envelope at the output of the EG combiner and the system's overall performance. This is done by considering three propagation scenarios called the full-LOS, the partial-LOS, and the NLOS scenario, denoted by $\mathrm{LOS}_{K,K}$, $\mathrm{LOS}_{K,0}$ ($\mathrm{LOS}_{0,K}$), and $\mathrm{LOS}_{0,0}$, respectively. Here, K corresponds to the number of mobile relays in the network. In the full LOS scenario, we have LOS components in the direct link as well as all the transmission links between the source mobile station and the destination mobile station via K mobile relays. The scenario in which LOS components are present in only a few links from the source mobile station to the destination mobile station via K mobile relays is referred to as the partial-LOS scenario. When LOS components do not exist in any of the transmission links, we have the NLOS scenario. Whenever, there exists a LOS component in any of the transmission links, its amplitude ρ_i is taken to be unity. It is necessary to keep in mind that there is a direct link between the source mobile station and the destination mobile station, in addition to the links via K mobile relays. Therefore, the total number of diversity branches available is $K + 1$. The presented results in Figs. 2–8 display a good fit of the approximated analytical and the exact simulation results.

Figure 2 demonstrates the theoretical approximation for the PDF $p_{\Xi_\rho}(x)$ of $\Xi_\rho(t)$ described in (17). This figure illustrates the PDF $p_{\Xi_\rho}(x)$ under full-LOS, partial-LOS, and NLOS propagation conditions considering a different number of mobile relays K. It is quite obvious from the figure that for any value of K, the presence of LOS components increases both the mean value and the variance of $\Xi_\rho(t)$. Furthermore, for the $\mathrm{LOS}_{K,K}$ scenario if $K = 1$, the PDF $p_{\Xi_\rho}(x)$ maps to the double Rice distribution as $\sigma_0^2 \to 0$, whereas $p_{\Xi_\rho}(x)$ reduces to the double Rayleigh distribution for the $\mathrm{LOS}_{0,0}$ scenario. Another important result is that the PDF $p_{\Xi_\rho}(x)$ of $\Xi_\rho(t)$ tends to a Gaussian distribution if K increases. This observation is in accordance with the central limit theorem [47]. A close agreement between the approximated theoretical and the exact simulation results confirms the correctness of our approximation.

The LCR $N_{\Xi_\rho}(r)$ of $\Xi_\rho(t)$ described by (24) is evaluated along with the exact simulation results in Fig. 3. This figure presents the LCR $N_{\Xi_\rho}(r)$ of $\Xi_\rho(t)$ corresponding to the $\mathrm{LOS}_{K,K}$, $\mathrm{LOS}_{K,0}$ ($\mathrm{LOS}_{0,K}$), and $\mathrm{LOS}_{0,0}$ scenarios considering a different number of mobile relays K in the system. It can be observed that in general, for any value of K, at low signal levels r, the LOS components facilitate in decreasing the LCR $N_{\Xi_\rho}(r)$. However, at high signal levels r, the presence of LOS components contributes towards an increase in $N_{\Xi_\rho}(r)$. These results also illustrate that for all the three considered propagation scenarios, at any signal level r, (24) closely approximates the exact simulation results if $K > 1$. This is in contrast to the case

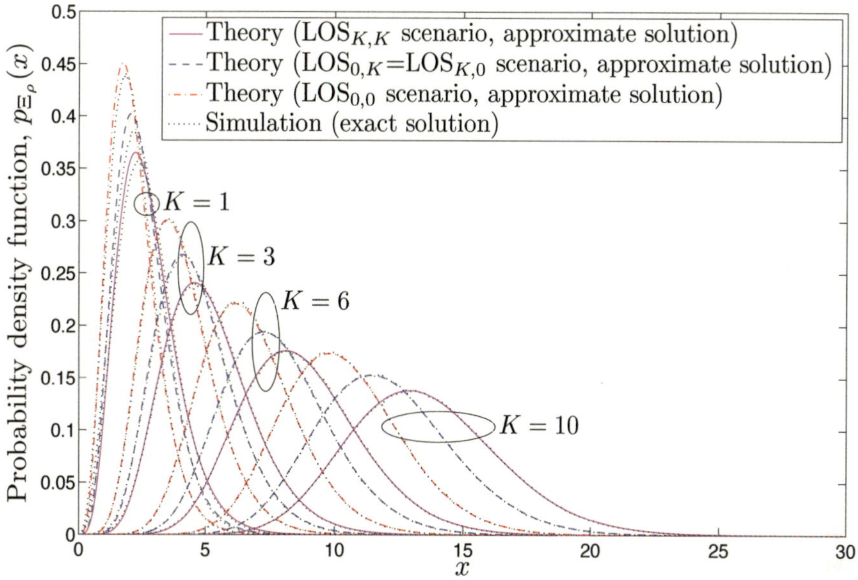

Figure 2. The PDF $p_{\Xi_\rho}(x)$ of the received signal envelope $\Xi_\rho(t)$ at the output of the EG combiner for $K+1$ diversity branches under different propagation conditions.

if $K = 1$, where (24) holds only for high values of r. We can further deduce from these results that by increasing K, $N_{\Xi_\rho}(r)$ reduces (increases) at low (high) values of r. It is also worth noticing that at high values of r, for the LOS$_{K,K}$ scenario if $K = 1$, as $\sigma_0^2 \to 0$, (24) provides us with a very close approximation to the exact LCR of a double Rice process given in [37], whereas it approximates well to the exact LCR of a double Rayleigh process for the LOS$_{0,0}$ scenario [56].

Figure 4 displays the analytical approximate results of the ADF $T_{\Xi_{\rho-}}(r)$ of $\Xi_\rho(t)$ described by (25) along with the exact simulation results. These results clearly indicate that for all propagation scenarios, i.e., the LOS$_{K,K}$, LOS$_{K,0}$ (LOS$_{0,K}$), and LOS$_{0,0}$ scenarios, an increase in the number K of mobile relays results in a decrease of $T_{\Xi_{\rho-}}(r)$ at all signal levels r. It can also be observed in Fig. 4 that the presence of the LOS components in all the transmission links lowers $T_{\Xi_{\rho-}}(r)$ for all signal levels r and any number K.

The average BEP P_b of M-ary PSK modulation schemes over M2M fading channels with LOS components and EGC described by (36) is presented in Fig. 5. In this figure, a comparison of the average BEP P_b of quadrature PSK (QPSK), 8-PSK, as well as 16-PSK modulation schemes is shown by taking into account $K+1$ diversity branches for each modulation scheme. The average BEP P_b curves associated with the aforementioned modulation schemes in double Rice channels are also included in Fig. 5. Here, the average BEP P_b is evaluated for the LOS$_{K,K}$ scenario, i.e., $\rho_i = 1\,\forall\, i = 0,1,2,\ldots,2K$. For all modulation schemes, if $K = 1$, a significant enhancement in the diversity gain can be observed with the availability of just one extra transmission link. See, e.g., if the direct link from the source mobile station to the destination mobile station is not blocked by obstacles, and if there is one relay present in the system, then it is possible to attain a diversity gain of approximately 21 dB at $P_b = 10^{-3}$. Increasing the number K of mobile relays in the system, in turn increases the number of diversity branches and hence improves the performance. The provision of higher data rates is the characteristic

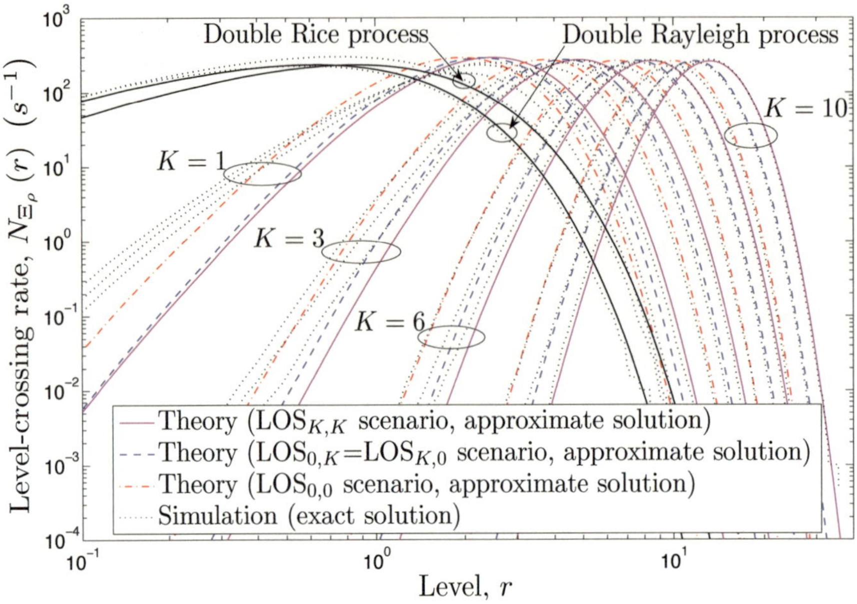

Figure 3. The LCR $N_{\Xi_\rho}(r)$ of the received signal envelope $\Xi_\rho(t)$ at the output of the EG combiner for $K+1$ diversity branches under different propagation conditions.

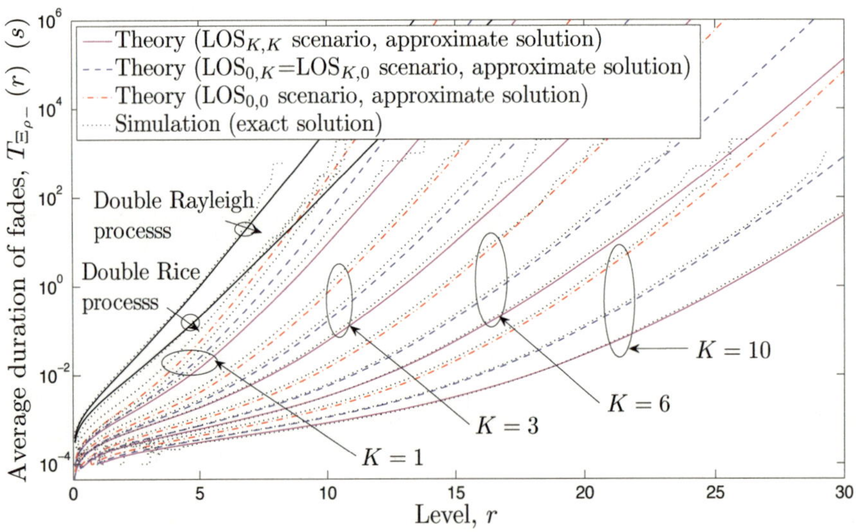

Figure 4. The ADF $T_{\Xi_{\rho-}}(r)$ of the received signal envelope $\Xi_\rho(t)$ at the output of the EG combiner for $K+1$ diversity branches under different propagation conditions.

feature of higher-order modulation schemes. These modulations are however known to be more prone to transmission errors. This sensitivity of higher-order modulations towards transmission errors is visible in Fig. 5 as the average BEP P_b curve associated with QPSK modulation shifts to the right if 8-PSK or 16-PSK modulation schemes are deployed.

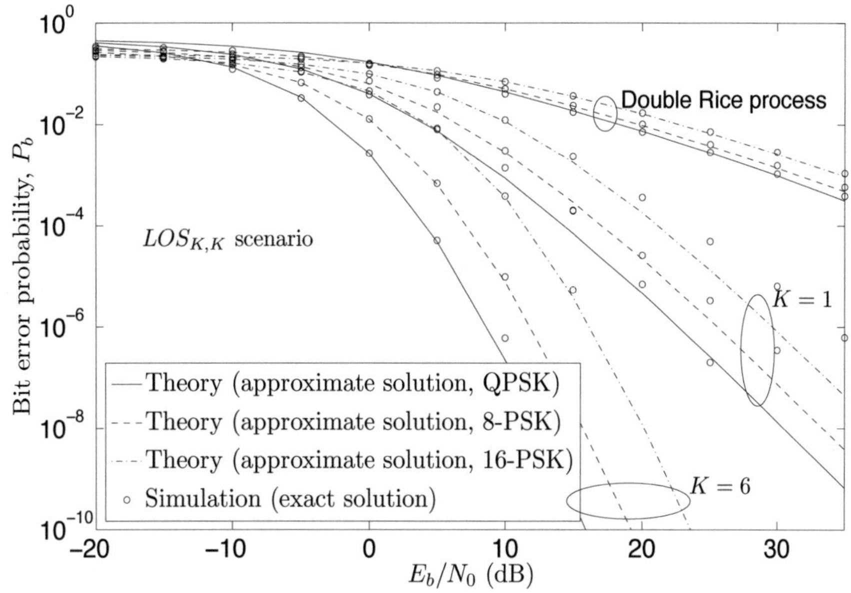

Figure 5. The average BEP P_b of M-ary PSK modulation schemes over M2M fading channels with EGC under full-LOS propagation conditions.

An EG combiner installed at the destination mobile station makes a receiver diversity system. In addition to the diversity gain, such systems offer an array gain as well [54]. The array gain in fact results from coherent combining of multiple received signals. In the context of EGC, the array gain allows the receiver diversity system in a fading channel to achieve a better performance than a system without diversity in an AWGN channel with the same average SNR [54]. Figure 6 includes the theoretical results of the average BEP P_b of QPSK under full-LOS propagation conditions (i.e., the $\text{LOS}_{K,K}$ scenario) with increasing number K of diversity branches. In the presented results, $K \geq 10$ implies that we have at least 11 diversity branches. Note that in Fig. 6, for $K \geq 10$, the dual-hop amplify-and-forward system with M2M fading channels has a lower error probability than a system in an AWGN channel with the same SNR. This improved performance is due to the array gain of the EG combiner.

Figure 7 illustrates the impact of the presence of LOS components in the relay links on the average BEP P_b of M-ary PSK modulation schemes. Keeping the number of diversity branches constant, e.g., for $K = 3$, the average BEP P_b of QPSK and 16-PSK modulation schemes is evaluated for the $\text{LOS}_{K,K}$, $\text{LOS}_{K,0}$ ($\text{LOS}_{0,K}$), and $\text{LOS}_{0,0}$ scenarios. For both QPSK and 16-PSK modulations, there is a noticeable gain in the performance if the scenario changes from $\text{LOS}_{0,0}$ to $\text{LOS}_{K,K}$. See, e.g., at $P_b = 10^{-4}$, a gain of approximately 1.5 dB is achieved when we have $\text{LOS}_{K,0}$ ($\text{LOS}_{0,K}$) compared to $\text{LOS}_{0,0}$. A further increase of approximately 1 dB in the gain can be seen if $\text{LOS}_{K,K}$ conditions are available.

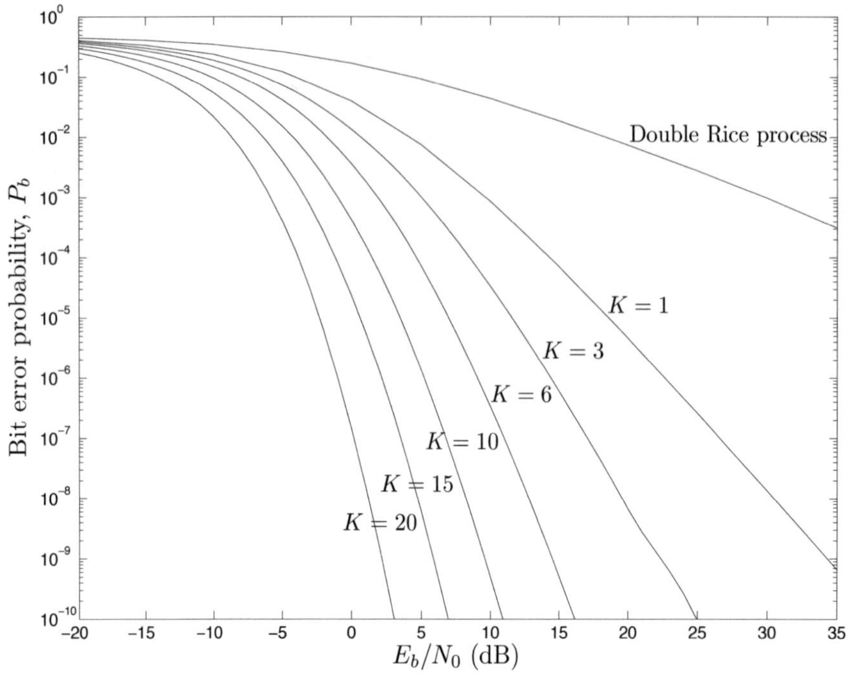

Figure 6. The average BEP P_b of QPSK modulation schemes over M2M fading channels with EGC for the $LOS_{K,K}$ scenario.

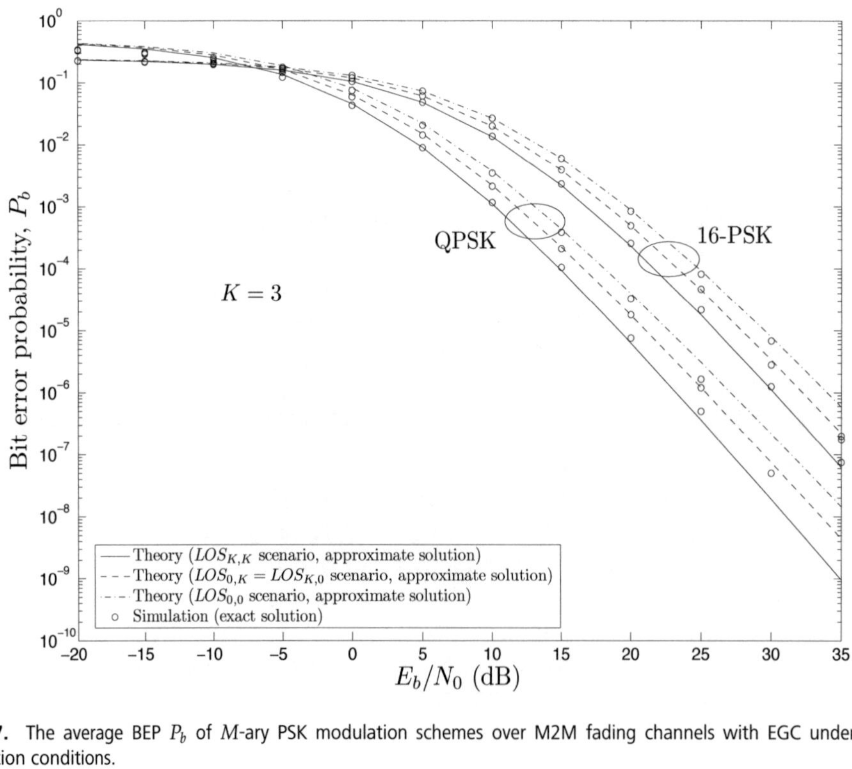

Figure 7. The average BEP P_b of M-ary PSK modulation schemes over M2M fading channels with EGC under different propagation conditions.

http://dx.doi.org/10.5772/55333

Finally, the outage probability $P_{\text{out}}(\gamma_{th})$ described by (37) is evaluated along with the exact simulation results in Fig. 8. Under full-LOS propagation conditions with the QPSK modulation scheme employed in our analysis, $P_{\text{out}}(\gamma_{th})$ is obtained for a different number of diversity branches. The presented results show a decrease in $P_{\text{out}}(\gamma_{th})$, which is due to EGC deployed at the destination mobile station, where the resulting performance advantage is the diversity gain.

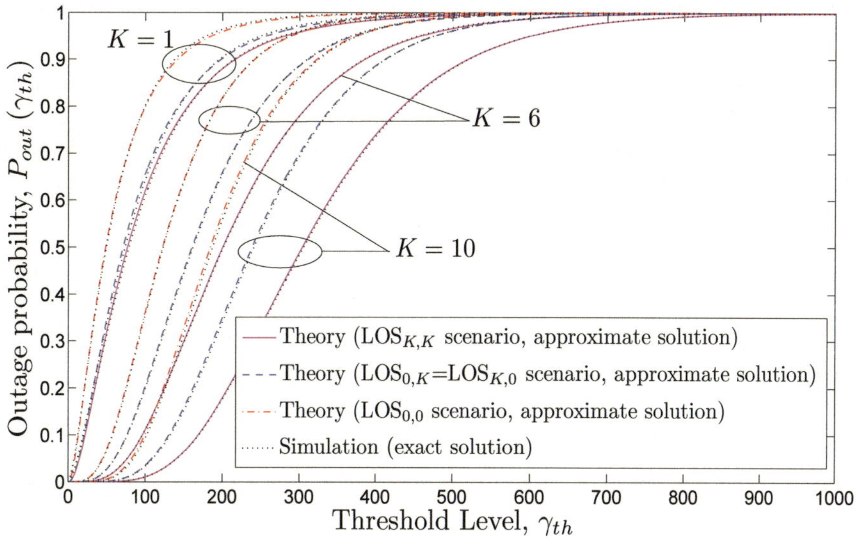

Figure 8. The outage probability $P_{\text{out}}(\gamma_{th})$ in M2M fading channels with EGC under different propagation conditions.

6. Conclusion

This article provides a profound study pertaining to the statistical properties of EGC over M2M fading channels under LOS propagation conditions in relay-based networks. In addition vital information about the performance of relay-based cooperative systems in such channels is made available. The system under investigation is a dual-hop amplify-and-forward relay communication system, where there exist K mobile relays between the source mobile station and the destination mobile station. It is further assumed that the direct link from the source mobile station to the destination mobile station is not blocked by any obstacles. Such a configuration gives rise to $K+1$ diversity branches. The signals received from the $K+1$ diversity branches are then combined at the destination mobile station to achieve the spatial diversity gain. In order to accommodate the direct link along with the unbalanced relay links, we have modeled the received signal envelope at the output of the EG combiner as a sum of a classical Rice process and K double Rice processes. Furthermore, the classical Rice process and double Rice processes are independent. Note that these double Rice processes are independent but not necessarily identically distributed.

The statistical analysis is carried out by deriving simple and closed-form analytical approximations for the channel statistics such as the PDF, CDF, LCR, and ADF. Here, the Laguerre series expansion has been employed to approximate the PDF of the sum of classical Rice and K double Rice processes. The advantage of using the Laguerre series is that this allows to approximate the PDF of the sum process by a gamma distribution with reasonable accuracy. The CDF, LCR, and ADF of the sum process are also approximated by exploiting the properties of a gamma distributed process. Furthermore, the presented results demonstrate that the approximated theoretical results fit closely to the exact simulation results. From this fact, we can conclude that the approximation approach outlined in this study is quite useful, general, and easy to implement. In addition to studying the impact of the number of diversity branches, we have included in our discussion the influence of the existence of the LOS components in the transmission links on the statistical properties of EGC over M2M channels.

The utilization of the presented statistical analysis is then demonstrated in the performance evaluation of dual-hop multi-relay cooperative systems. In this work, the performance assessment measures of interest are the PDF as well as the moments of the SNR, AOF, the average BEP, and the outage probability. The PDF of the SNR is obtained from the previously derived PDF of the sum process by a simple transformation of random variables. Starting from the PDF of the SNR, the computation of the moments of the SNR, AOF, and the outage probability is rather straightforward.

Author details

Batool Talha and Matthias Pätzold

Department of Information and Communication Technology, Faculty of Engineering and Science, University of Agder, Grimstad, Norway

References

[1] A. Sendonaris, E. Erkip, and B. Aazhang. User cooperation diversity — Part I: System description. *IEEE Trans. Commun.*, 51(11):1927–1938, November 2003.

[2] A. Sendonaris, E. Erkip, and B. Aazhang. User cooperation diversity — Part II: Implementation aspects and performance analysis. *IEEE Trans. Commun.*, 51(11):1939–1948, November 2003.

[3] J. N. Laneman, D. N. C. Tse, and G. W. Wornell. Cooperative diversity in wireless networks: Efficient protocols and outage behavior. *IEEE Trans. Inform. Theory*, 50(12):3062–3080, December 2004.

[4] R. Pabst, B. Walke, D. Schultz, et al. Relay-based deployment concepts for wireless and mobile broadband radio. *IEEE Communications Magazine*, 42(9):80–89, September 2004.

[5] Y. Fan and J. S. Thompson. MIMO configurations for relay channels: Theory and practice. *IEEE Trans. Wireless Commun.*, 6(5):1774–1786, May 2007.

[6] M. Dohler and Y. Li. *Cooperative Communications: Hardware, Channel, and PHY*. Chichester: John Wiley & Sons, 1st edition, 2010.

[7] W. C. Jakes, editor. *Microwave Mobile Communications*. Piscataway, NJ: IEEE Press, 1994.

[8] M. D. Yacoub, C. R. C. Monterio da Silva, and J. E. V. Bautista. Second-order statistics for diversity-combining techniques in Nakagami fading channels. *IEEE Trans. Veh. Technol.*, 50(6):1464–1470, November 2001.

[9] N. C. Beaulieu and X. Dong. Level crossing rate and average fade duration of MRC and EGC diversity in Ricean fading. *IEEE Trans. Commun.*, 51(5):722–726, May 2003.

[10] P. Ivanis, D. Drajic, and B. Vucetic. The second order statistics of maximal ratio combining with unbalanced branches. *IEEE Communications Letters*, 12(7):508–510, July 2008.

[11] Q. T. Zhang. Probability of error for equal-gain combiners over Rayleigh channels: some closed-form solutions. *IEEE Trans. Commun.*, 45(3):270–273, March 1997.

[12] D. A. Zogas, G. K. Karagiannidis, and S. A. Kotsopoulos. Equal gain combining over Nakagami-n (Rice) and Nakagami-q (Hoyt) generalized fading channels. *IEEE Trans. Wireless Commun.*, 4(2):374–379, March 2005.

[13] H. Samimi and P. Azmi. An approximate analytical framework for performance analysis of equal gain combining technique over independent Nakagami, Rician and Weibull fading channels. *Wireless Personal Communications (WPC)*, 43(4):1399–1408, dec 2007. DOI 10.1007/s11277-007-9314-z.

[14] O. M. Hasna and M. S. Alouini. End-to-end performance of transmission systems with relays over Rayleigh-fading channels. *IEEE Trans. Wireless Commun.*, 2(6):1126 - 1131, November 2003.

[15] P. A. Anghel and M. Kaveh. Exact symbol error probability of a cooperative network in a Rayleigh-fading environment. *IEEE Trans. Wireless Commun.*, 3(5):1416–1421, September 2004.

[16] L. Wu, J. Lin, K. Niu, and Z. He. Performance of dual-hop transmissions with fixed gain relays over generalized-*K* fading channels. In *Proc. IEEE Int. Conf. Communications (ICC'09)*. Dresden, Germany, June 2009. DOI 10.1109/ICC.2009.5199331.

[17] T. A. Tsiftsis, G. K. Karagiannidis, and S. A. Kotsopoulos. Dual-hop wireless communications with combined gain relays. *IEE Proc. Communications*, 152(5):528 – 532, October 2005.

[18] A. Ribeiro, X. Cai, and G. B. Giannakis. Symbol error probabilities for general cooperative links. *IEEE Trans. Wireless Commun.*, 4(3):1264–1273, May 2005.

[19] Y. Li and S. Kishore. Asymptotic analysis of amplify-and-forward relaying in Nakagami fading environments. *IEEE Trans. Wireless Commun.*, 6(12):4256–4262, December 2007.

[20] W. Su, A. K. Sadek, and K. J. R. Liu. Cooperative communication protocols in wireless networks: Performance analysis and optimum power allocation. *Wireless Personal Communications (WPC)*, 44(2):181–217, January 2008.

[21] M. Di Renzo, F. Graziosi, and F. Santucci. A comprehensive framework for performance analysis of dual-hop cooperative wireless systems with fixed-gain relays over generalized fading channels. *IEEE Trans. Wireless Commun.*, 8(10):5060–5074, October 2009.

[22] H. Suraweera, G. Karagiannidis, and P. Smith. Performance analysis of the dual-hop asymmetric fading channel. *IEEE Trans. Wireless Commun.*, 8(6):2783 – 2788, June 2009.

[23] F. Xu, F. C. M. Lau, and D. W. Yue. Diversity order for amplify-and-forward dual-hop systems with fixed-gain relay under Nakagami fading channels. *IEEE Trans. Wireless Commun.*, 9(1):92 – 98, January 2010.

[24] D. B. da Costa and S. Aissa. Performance of cooperative diversity networks: Analysis of amplify-and-forward relaying under equal-gain and maximal-ratio combining. In *Proc. IEEE Int. Conf. Communications (ICC'09)*, pages 1–5. Dresden, Germany, June 2009. DOI 10.1109/ICC.2009.5199330.

[25] S. S. Ikki and M. H. Ahmed. Performance of cooperative diversity using equal gain combining (EGC) over Nakagami-*m* fading channels. *IEEE Trans. Wireless Commun.*, 8(2):557–562, February 2009.

[26] H. Q. Huynh, S. I. Husain, J. Yuan, A. Razi, and D. S. Taubman. Performance analysis of multi-branch non-regenerative relay systems with EGC in Nakagami-*m* channels. In *Proc. IEEE 70th Veh. Technol. Conf., VTC'09-Fall*, pages 1–5. Anchorage, AK, USA, September 2009. DOI 10.1109/VETECF.2009.5378710.

[27] W. R. Young. Comparison of mobile radio transmission at 150, 450, 900, and 3700 MHz. *Bell Syst. Tech. J.*, 31:1068–1085, November 1952.

[28] H. W. Nylund. Characteristics of small-area signal fading on mobile circuits in the 150 MHz band. *IEEE Trans. Veh. Technol.*, 17:24–30, October 1968.

[29] Y. Okumura, E. Ohmori, T. Kawano, and K. Fukuda. Field strength and its variability in VHF and UHF land mobile radio services. *Rev. Elec. Commun. Lab.*, 16:825–873, September/October 1968.

[30] M. Pätzold, U. Killat, and F. Laue. An extended Suzuki model for land mobile satellite channels and its statistical properties. *IEEE Trans. Veh. Technol.*, 47(2):617–630, May 1998.

[31] H. Suzuki. A statistical model for urban radio propagation. *IEEE Trans. Commun.*, 25(7):673–680, July 1977.

[32] S. Chitroub, A. Houacine, and B. Sansal. Statistical characterisation and modelling of SAR images. *Elsevier, Signal Processing*, 82(1):69–92, January 2002. DOI http://dx.doi.org/10.1016/S0165-1684(01)00158-X.

[33] P. M. Shankar. Error rates in generalized shadowed fading channels. *Wireless Personal Communications (WPC)*, 28(3):233–238, February 2004. DOI http://dx.doi.org/10.1023/B:wire.0000032253.68423.86.

http://dx.doi.org/10.5772/55333

[34] I. Z. Kovacs, P. C. F. Eggers, K. Olesen, and L. G. Petersen. Investigations of outdoor-to-indoor mobile-to-mobile radio communication channels. In *Proc. IEEE 56th Veh. Technol. Conf., VTC'02-Fall*, volume 1, pages 430–434. Vancouver BC, Canada, September 2002.

[35] M. Pätzold, B. O. Hogstad, and N. Youssef. Modeling, analysis, and simulation of MIMO mobile-to-mobile fading channels. *IEEE Trans. Wireless Commun.*, 7(2):510–520, February 2008.

[36] G. K. Karagiannidis, N. C. Sagias, and P. T. Mathiopoulos. $N{*}$Nakagami: A novel stochastic model for cascaded fading channels. *IEEE Trans. Commun.*, 55(8):1453–1458, August 2007.

[37] B. Talha and M. Pätzold. On the statistical properties of double Rice channels. In *Proc. 10th Int. Symp. on Wireless Personal Multimedia Communications, WPMC 2007*, pages 517âĂŞ–522. Jaipur, India, December 2007.

[38] H. Ilhan, M. Uysal, and I. Altunbas. Cooperative diversity for intervehicular communication: Performance analysis and optimization. *IEEE Trans. Veh. Technol.*, 58(7):3301 – 3310, August 2009.

[39] W. Wongtrairat and P. Supnithi. Performance of digital modulation in double Nakagami-m fading channels with MRC diversity. *IEICE Trans. Commun.*, E92-B(2):559–566, February 2009.

[40] B. Talha, M. Pätzold, and S. Primak. Performance analysis of M-ary psk modulation schemes over multiple double Rayleigh fading channels with EGC in cooperative networks. In *Proc. IEEE ICC 2010 Workshop on Vehicular Connectivity*, pages 1–6. Cape Town, South Africa, May 2010. DOI: 10.1109/ICCW.2010.5503940.

[41] R. U. Nabar, H. Bölcskei, and F. W. Kneubühler. Fading relay channels: Performance limits and space-time signal design. *IEEE J. Select. Areas Commun.*, 22(6):1099–1109, August 2004.

[42] K. Azarian, H. E. Gamal, and P. Schniter. On the achievable diversity-multiplexing tradeoff in half-duplex cooperative channels. *IEEE Trans. Inform. Theory*, 51(12):4152–4172, December 2005.

[43] Y. Wu and M. Pätzold. Parameter optimization for amplify-and-forward relaying systems with pilot symbol assisted modulation scheme. *Wireless Sensor Networks (WSN)*, 1(1):15âĂŞ–21, April 2009. DOI 10.4236/wsn.2009.11003.

[44] S. Primak, V. Kontorovich, and V. Lyandres, editors. *Stochastic Methods and their Applications to Communications: Stochastic Differential Equations Approach*. Chichester: John Wiley & Sons, 2004.

[45] I. S. Gradshteyn and I. M. Ryzhik. *Table of Integrals, Series, and Products*. New York: Academic Press, 6th edition, 2000.

[46] M. Pätzold. *Mobile Fading Channels*. Chichester: John Wiley & Sons, 2002.

[47] A. Papoulis and S. U. Pillai. *Probability, Random Variables and Stochastic Processes.* New York: McGraw-Hill, 4th edition, 2002.

[48] S. O. Rice. Mathematical analysis of random noise. *Bell Syst. Tech. J.*, 24:46–156, January 1945.

[49] M. Nakagami. The *m*-distribution: A general formula of intensity distribution of rapid fading. In W. G. Hoffman, editor, *Statistical Methods in Radio Wave Propagation.* Oxford, UK: Pergamon Press, 1960.

[50] M. D. Yacoub, J. E. V. Bautista, and L. G. de Rezende Guedes. On higher order statistics of the Nakagami-*m* distribution. *IEEE Trans. Veh. Technol.*, 48(3):790–794, May 1999.

[51] M. K. Simon and M. S. Alouini. *Digital Communications over Fading Channels.* New Jersey: John Wiley & Sons, 2nd edition, 2005.

[52] M. Schwartz, W. R. Bennett, and S. Stein. *Communication Systems and Techniques*, volume 4. New York: McGram Hill, 1966.

[53] U. Charash. Reception through Nakagami fading multipath channels with random delays. *IEEE Trans. Commun.*, 27(4):657–670, April 1979.

[54] A. Goldsmith. *Wireless Communications.* New York: Cambridge University Press, 2005.

[55] M. Pätzold, C. X. Wang, and B. O. Hogstad. Two new sum-of-sinusoids-based methods for the efficient generation of multiple uncorrelated Rayleigh fading waveforms. *IEEE Trans. Wireless Commun.*, 8(6):3122âĂŞ–3131, June 2009.

[56] C. S. Patel, G. L. Stüber, and T. G. Pratt. Statistical properties of amplify and forward relay fading channels. *IEEE Trans. Veh. Technol.*, 55(1):1–9, January 2006.